AF588108

Microsystems Dynamics

International Series on
INTELLIGENT SYSTEMS, CONTROL, AND AUTOMATION: SCIENCE AND ENGINEERING

VOLUME 44

For other titles published in this series, go to
www.springer.com/series/6259

Vytautas Ostasevicius • Rolanas Dauksevicius

Microsystems Dynamics

Vytautas Ostasevicius
Kaunas University of Technology
Institute for Hi-Tech Development
Studentu Street 65
LT 51369 Kaunas
Lithuania
vytautas.ostasevicius@ktu.lt

Rolanas Dauksevicius
Kaunas University of Technology
Institute for Hi-Tech Development
Studentu Street 65
LT 51369 Kaunas
Lithuania
rolanas.dauksevicius@ktu.lt

ISBN 978-90-481-9700-2 e-ISBN 978-90-481-9701-9
DOI 10.1007/978-90-481-9701-9
Springer Dordrecht Heidelberg London New York

Library of Congress Control Number: 2010937993

Printed on acid-free paper

Springer is part of Springer Science+Business Media (www.springer.com)

Preface

Elastic- and rotary-type microelectromechanical systems (MEMS) constitute a significant portion of current microsystems technology. Microcantilever-based contact-type sensors and actuators as well as microrotor-based devices have a huge application potential in different fields of industry including telecom, IT, material characterization, chemistry and environmental monitoring, (bio)medical, automotive, commercial as well as aerospace and defense applications.

This research monograph focuses on the dynamic aspects of elastic vibro-impact microsystems by presenting an in-depth numerical analysis of novel effects of structural dynamics together with optimization results of microcantilevers, which are investigated purely from mechanical point of view, thereby concentrating on the complex impact process and the intrinsic dynamic properties of elastic structures such as natural vibration modes and their advantageous utilization. Furthermore, the book addresses important problems in MEMS dynamics related to behavior of microstructures under the influence of electrostatic actuation and viscous air damping. Computational models are proposed that take into account interactions between a microcantilever and forces generated from the surrounding air, electrostatic field and impacts with a rigid surface including microscale-dominant adhesive interactions. The authors also provide coverage of key aspects of the design and control of electrostatic micromotors by presenting corresponding computational models and major findings of numerical analysis. Finally, the monograph presents design and fabrication technology of the developed electrostatic microswitches and micromotors including also experimental results with particular attention on vibration testing of MEMS structures by means of laser Doppler vibrometry.

Microsystems dynamics provides a detailed treatment of dynamic phenomena, which are important to MEMS field and therefore serves as a guide for designers involved in R&D of advanced microsystems that are based on operation of elastic and/or rotary microstructures.

Kaunas, Lithuania
January 2010

Vytautas Ostasevicius
Rolanas Dauksevicius

Contents

Chapter 1
Introduction

Abstract The first chapter introduces to the field of microelectromechanical systems (MEMS) by providing a concise coverage of microfabrication technologies, common types of microactuators with particular emphasis on electrostatic MEMS switches and motors. Classification of microswitches is provided together with major application areas of micromotors. The chapter is concluded with pointing out the importance of research of MEMS dynamics and briefly outlining the scope of the monograph.

1.1 Review of Microelectromechanical Systems (MEMS)

Technology of microelectromechanical systems (MEMS) is at the center of a rapidly emerging industry combining many different engineering disciplines and physics: electrical, electronic, mechanical, optical, material, chemical, and fluidic engineering. In its most conventional sense MEMS refers to a class of batch-fabricated devices that utilize both mechanical and electrical components to simulate macroscopic devices on a microscopic scale. Over the past 30 years, there has been tremendous investment in the microelectronics infrastructure worldwide. MEMS industry began as a successor of integrated circuit industry using often the same manufacturing and testing equipment adapted to its own needs. In 1958, J. Kilby from company Texas Instruments created the first electronic circuit in which all the components were fabricated in a single piece of semiconductor material 17 mm^2 in size. But when he pressed the switch, the first integrated circuit (IC) worked, showing an unending sine curve across an oscilloscope screen. A few years later R. Moyce (who co-founded Intel with G. Moore in 1968) developed the first industrial IC-fabrication process transforming former room-size machines into today's personal computers. Since early sixties, the growth in chip performance has obeyed the Moore's Law, which states that the number of transistors on an

V. Ostasevicius and R. Dauksevicius, *Microsystems Dynamics*,
Intelligent Systems, Control and Automation: Science and Engineering 44,
DOI 10.1007/978-90-481-9701-9_1,

integrated circuit for minimum component cost doubles every 24 months. MEMS is the next logical step after ICs that began over three decades ago. Less than 10 years after the invention of the integrated circuit, H. C. Nathanson used 14 microelectronic fabrication techniques to make the world's first micromechanical device [1]. It is generally agreed that the first MEMS device was his gold resonating MOS gate structure. As the smallest commercially produced "machines", MEMS devices are similar to traditional sensors and actuators although much smaller. Complete systems are typically a few millimeters across, with individual features/devices of the order of 1–100 μm across. The miniaturization and integration example, successfully pioneered in the microelectronics industry, has been gradually extended to domains other than electronics. MEMS devices can be produced using sophisticated deposition tools, lithographic processes, wire bonders, photoresists and silicon wafers. The worldwide market of MEMS is estimated to exceed tens of billions of dollars in the next several years. Just several years ago successful mass produced micromachined devices of commercial importance were mainly automotive pressure sensors, airbag accelerometers, automotive fuel injection nozzles and printer ink spraying nozzles. Variations of MEMS sensor and actuator technology are used to build micromotors. Recent advances in technology have put many of these sensor processes on exponentially decreasing size/power/cost curves. The most *obvious* characteristic of MEMS is that they are small. But the most *important* characteristic of MEMS is that they can be made cheaply. If the Pentium chip itself occupied a volume of 100 times its present size, with its other performance and price characteristics the same, the overall impact of integrated circuits on the world would not be dramatically affected. On the other hand, if Pentium processors were 100 times as expensive, all of the sudden, many applications, where they are presently used, could no longer afford them. Their overall impact on the world would be significantly less. Hence, the real impact of integrated circuits, and analogously MEMS, does not come from their size, but from their cost. Offering economies unique to the semiconductor industry, MEMS are revolutionizing the industrial age. The effects of MEMS are enacting sweeping reforms within the space industry. NASA hopes to eventually phase out the large satellites that it employs to reach the farthest points in the solar system. With every kilogram sent to Mars costing upwards of one million dollars, the potential of sending a fully integrated spacecraft weighting a few kilograms instead of the thousands of kilos offers significant monetary benefits [1]. While the growth in the functionality of microelectronic circuits has been truly phenomenal, for the most part this growth has been limited to data processing, storage and communication. The power and capability of the chips of tomorrow will transcend simply storing and processing information. The hallmark of the next 30 years will be the incorporation of new types of functionality onto the chip; structures that will enable the chip to not only reason, but to sense, act and communicate as well. This will result in chips that will have an ability to affect their surroundings and to communicate to the outside world in new and dramatic ways.

1.2 MEMS Production Technologies

Tool-making for mechanical machining began perhaps a million years ago when humans learned to walk erect and had hands free to grasp objects. Since that time, humans have been cutting metal to make many tools, machines and other devices that our civilization demands. MEMS can be produced in two fundamentally distinct ways:

(a) The "top-down" approach, which makes use of historical tool-machining techniques (turning, milling, drilling, etc.) by downsizing conventional mechanisms
(b) The "bottom-up" approach, which makes use of revolutionary mask-machining techniques, combining many elementary components duplicated within a parallel process flow

Top-down approach. Today, there are many research groups that aim at fabrication of small devices using conventional techniques such as cutting, drilling, sand blasting, electrodischarge machining (EDM), ultrasonic machining (USM), mold injection, etc. Ultrahigh precision machines with sharp single-crystal diamond tools have already made sub-micrometer precision machining possible. Similar manufacturing accuracy has been also obtained using USM and EDM techniques [2]. Finest details in rather complicated three-dimensional structures can be produced using conventional tool-machining technology in the range of 10–100 μm, also. The main limitation of traditional machining approaches is that using conventional machining technology, the components of a system must be made piece by piece. So, even if conventional precision mechanics satisfied most of current miniaturization constraints, the cost-effective parallel integration of MEMS cannot be achieved using mass machining approach.

Bottom-up approach. Bulk and surface micromachining are the most popular and most promising bottom-up approach techniques, though they can provide only quasi-three dimensionality in structures. Surface lithography gives high resolution in plane, but due to processing parameters such as under-etching and etch uniformity over a wafer, the resolution in the direction normal to the plane is much poorer. Thus, in general, conventional photolithography techniques limit structures only to two dimensions.

1.3 Common MEMS Actuators

An interesting difference between macroscopic and microscopic devices is the nature of forces that move things around. At macroscopic scale, electromagnetic forces tend to be used to build strong motors. Magnets and coils of wire with currents in them can produce large forces. Here, electrostatic forces are mainly useful for amusements like sticking balloons to a wall. At the microscale, electromagnetic forces get proportionally smaller when sizes of wires and magnets scale down, while electrostatic forces become large enough to be practical as objects get closer together. When dimensions of objects become smaller than some millimeters, the relative forces generated by electromagnetic fields abruptly decrease until become nearly impossible to use. On the other hand, electrostatic force becomes more and more useful for various actuation applications. This is why electrostatic force is often regarded as the main drive source for microactuators. There are three basic types of actuation and sensing mechanisms that use electrostatic force: comb drives, parallel-plate capacitors (e.g. microswitches) and micromotors. The interest of this monograph lies in the last two categories.

1.4 RF MEMS and Microswitches

MEMS for radio frequency (RF) applications or RF MEMS is one of the fastest growing areas in commercial MEMS technology that promises breakthrough advances in telecommunications, radar systems, personal mobiles, etc. RF MEMS refers to the application of MEMS technology to high-frequency circuits (i.e. RF, microwave, or millimeter wave). Compared to other more mature MEMS fields, RF MEMS is relatively new but has already generated a tremendous amount of excitement because of both performance enhancement and manufacturing cost reduction [3, 4]. RF MEMS is considered a priority research area in Europe and therefore under the EU Framework VI programme, a fully-funded Network of Excellence (NoE) in RF MEMS and RF Microsystems was created. Called "Advanced MEMS for RF and Millimeter Wave Communications" (AMICOM), this 3-year NoE was officially launched on 1 January 2004. AMICOM is made up of a virtual network of 25 research institutes, across 14 countries, all working in RF MEMS technologies [5]. RF MEMS consist of passive components, such as transmission lines and couplers, and active components, such as variable capacitors, filters and *microswitches*. The latter are one of the most important active RF MEMS components. MEMS switch has been first conceptualized and fabricated by K. Petersen at IBM in 1979 [6]. It was produced on silicon, with an electrostatically actuated SiO_2 membrane. But these microdevices remained an exotic curiosity until 1990s, when the first microswitch for microwave applications was demonstrated [7]. Interest in MEMS switches has surged recently, principally due to the demonstrated performance benefits in switching RF signals. Currently electronic switches (diodes and transistors) and conventional electromagnetic relays are used to perform switching function, however low-power handling and high resistive losses are characteristic to semiconductor switches, and, in contrast electromagnetic relays are high-power devices, but useful only at lower radio frequencies, and operate at a much lower speed. Thus MEMS technology enabled to develop microswitches that overcame most of the disadvantages of current switching devices and combined their best advantages: ultra-high linearity, low insertion loss and power consumption (typically 0.05–0.1 mW), small size (<0.1 mm^2), wider frequency range (DC-40 GHz), good isolation, relatively low cost and ability to combine with electronics into "system-on-a-chip". Application areas of microswitches are numerous: automated test equipment, telecommunication systems (satellite, wireless), radar systems for automotive and defense applications, different industrial and medical equipment. The North American relay market alone is US $1.4 billion, and the growth in market share of surface-mount relays, from 1.2% in 1993 to 19.6% in 1997, demonstrates the value of miniaturization [8]. Until these days a large number of research institutions (e.g. University of California-Berkeley, Massachusetts Institute of Technology (MIT), National Taiwan University, Sandia labs) and private companies (e.g. Sony, Motorola, LG, NEC, Samsung, Bosch, Daimler-Chrysler, Raytheon, Analog Devices, Siemens) have built MEMS switches. However, their wider commercial availability is still hindered today because of: (a) challenges and high cost related to packaging – hermetic sealing is essential since microswitches are sensitive to humidity. Wafer-level packaging is the cheapest solution therefore intensive research is currently underway in order to find suitable bonding method; (b) drawbacks related to operational characteristics: higher actuation voltages (typically 20–80 V), lower switching speed (typically 1–300 μs), limited power handling capabilities (<1 W), and insufficient long-term reliability (lifetime: $\sim 10^{12}$ cycles testing mechanically (cold switching), $\sim 10^{9}$ – testing electrically (hot switching)). Thus modeling and experimental research of MEMS switches is essential for design optimization and device reliability.

Classification of microswitches. There are two distinct parts in the RF MEMS switch: the actuation (mechanical) section and the electrical section. The forces required for the mechanical movement can be obtained using electrostatic, magnetostatic, piezoelectric, or thermal actuation. Electrostatic actuation is the most prevalent technique in use today due to its virtually zero power consumption, small electrode size, thin layers used, and relatively short switching time. The switches can move either vertically or laterally, depending on their layout. The switching element is usually a cantilever beam or fixed-fixed beam. As regards electrical part, MEMS switches can be placed in either serial or shunt circuit configurations. As for contact type, these switches can be: (a) resistive (metal–metal). In a resistive microswitch (also referred to as *microrelay*) upon application of actuation voltage the tip of the beam approaches and collapses on the metallic bottom electrode thus achieving ohmic contact, closing an external circuit and allowing the current to flow, (b) capacitive (metal–insulator–metal), in which, upon actuation the beam (bridge) bends and collapses on the dielectric layer, changing the capacitance of an external circuit and the corresponding current intensity. As the actuation voltage is removed, the elastic restoring force returns the beam to its initial position thus opening the external circuit.

1.5 Applications of MEMS Micromotors

Not all MEMS prototypes have found their industrial use yet. Micromotors are one of them. The possible advantages of electrostatic micromotors are that they are very small, but currently the major disadvantages are that they produce very low torque and have high wear. Nevertheless, the most promising applications of micromotors are:

(a) Control: single unit navigation systems and control in distributed aerodynamic and hydrodynamic systems as microgyroscopes.
(b) Medical: miniature biochemical analytical instruments and drug delivery systems as micro servomotors and drives for micropumps, also cardiac management systems (e.g., pacemakers, catheters) as micromotors for probes.
(c) Optical: electromechanical signal processing as rotational micro switches.

MEMS gyroscopes. With dramatically reduced cost, size, and weight, as compared to current gyroscopes, MEMS gyroscopes potentially have a wide application spectrum in the aerospace industry, military, automotive industry and consumer electronics market. The automotive industry applications are diverse, including high performance navigation and guidance systems, ride stabilization, advanced automotive safety systems like yaw and tilt control, roll-over detection and prevention, and next generation airbag and anti-lock brake systems. A very wide range of consumer electronics applications include image stabilization in video cameras, virtual reality products, inertial pointing devices, and computer gaming industry. Miniaturization of gyroscopes also enables higher-end applications including micro-satellites, micro-robotics, and even implantable devices to cure vestibular disorders. Gyroscopes play an important role in the inertial navigation systems of flight vehicles, and promise several applications in automobile, defense, consumer electronics, and biomedical engineering. Major performance requirements of these applications are rate-grade resolution and bias stability, both being less than 0.5°/s [9]. The classical rotating wheel, precision fiber optic and ring laser gyroscopes are usually heavy, large and costly for most emerging applications. In contrast, microgyroscopes are light, small in size and potentially cheap. However, the performance of microgyroscopes has not yet approached that of the classical spinning mass gyroscopes.

Medical applications. The predicted advance of MEMS technology into medical market is still at its beginnings. One of the reasons is that the market for microengineered medical products is extremely competitive. Companies therefore mainly rely on existing technologies, which are highly developed, instead of investing in production facilities for microcomponents. The demand on miniaturized medical commercial products concentrates on millimeter sized devices. Micromotor can be an important part for a number of applications in medical technology if it can offer sufficient torque. Most of these applications do not need a 'true' micromotor with a diameter of 100 μm or less, but require a miniaturized drive with outer dimensions in the range of some millimeter delivering torque of typically 0.1–1 μNm [10]. Ultrasonic catheter systems have already been constructed making use of a LIGA-geared synchronous micromotor at the catheter tip to drive a transducer, though the motion is provided by piezo-transducers [11]. Current radial scanning systems use a motor from outside the body to drive a flexible shaft running through a long catheter tube to rotate a transducer at the catheter tip. However, that results in non-uniform rotational distortions, degradation of resolution and geometric accuracy, also smaller bending radii of the catheter are not possible. A sufficient torque micromotor integrated in the tool would solve many of these problems and make itself a very favorable product for medical markets. Moreover, nearly all products in medicine are of onetime use, thus high wear of micromotor is not critically important as long as the lifetime of a micromotor is enough to last one surgical operation.

Optical applications. MEMS structures in the micro- to millimeter range whose purposes are to manipulate light are often referred as micro-opto-electromechanical systems (MOEMS). Most often such device redirects chosen wavelengths of incoming light to a set of predefined positions. Such optical switch routes data from one set of optical fibers to another at locations called nodes. With trillions of bits of data per second on a single fiber and hundreds to thousands of fibers entering and leaving a large node, the aggregate switching capacity can be enormous. Currently the standard way of switching data is to convert the optical signals into electrical ones, use a large, fast electronic switch to route them, and then turn the signals back into light for transmission. This technique creates an electronic bottleneck. The ability of fibers to transmit optical data has outstripped the ability of electronic devices to convert that data into electrical signals and switch it. The basic idea of MEMS devices such as micromotors used for switches is to use microscopic mirrors to direct beams of light from many inputs to many outputs without slowing down the data streams. Rerouting light with MOEMS rotary switches not only breaks the electronic bottleneck, but is data rate independent, they can be fast, small, use little power, and above all, are inexpensive.

1.6 Importance of Dynamics

MEMS community generally agrees that structural dynamics is of key importance for successful realization of microsystems. It provides both vast opportunities and grand challenges for researchers in this field. There are still many dynamics-related issues that need to be tackled during design of microsystems. Designing a commercially viable MEMS device with required dynamic performance is possible only by means of thorough understanding and accurate prediction of its dynamic characteristics via modeling and simulation that are considered to be one of the most demanding tasks during MEMS development. Majority of MEMS components are transducers by their nature and include many physical energy domains such as mechanical, electrical, fluidic, thermal, etc. Therefore operation of microsystems involves complicated interaction of forces of different physical nature. In some cases MEMS components are relatively simple from structural point of view (e.g. beams, membranes, plates) however, usually they exhibit high functional complexity due to sophisticated physical operating principles. In many cases these interactions are intrinsically nonlinear except for small-amplitude motion of microstructures around equilibrium position. Thus major modeling difficulty of MEMS is due to the interdependency of several energy domains [12]. As a result, a researcher has to deal with so-called coupled-field systems and associated coupled-field (also referred to as coupled-domain or multiphysics) problems (e.g. problems involving fluidic-structural, electrostatic-structural and other interactions). Coupled-field systems are defined as systems that have several energy domains interacting with each other with the independent solution of any one domain being impossible without simultaneous solution of the others [13]. In coupled-field problems there are significant differences in temporal and spatial scales, which: (a) means that multiple grids and heterogeneous time-stepping algorithms may be needed for discretization, leading to very complicated and consequently computationally prohibitive simulation algorithms; (b) causes problems with numerical stability; (c) requires new physical assumptions and approximations to be made for some of the components and physical domains [12].

The field of structural dynamics in general is very wide so we can focus only on several important areas, which are directly relevant to the dynamics of microsystems, especially to contact-type microdevices (such as microswitches) and rotary-type microsystems (such as micromotors), which have been co-developed by our research group in cooperation with different research partners. In particular, our emphasis here will be placed on numerical analysis of fluidic-structural, electrostatic-structural as well as vibro-impact interactions and the associated dynamic processes. This monograph presents new research results related to structural dynamics of elastic and rotary microsystems as well as summarizes some of the research results that have already been published by the authors on the subject matter over the past several years [14–36].

References

1. Stark B (ed) (1999) MEMS reliability assurance guidelines for space applications. JPL NASA publication, Pasadena, CA, pp 99–1
2. Preumont A (ed) (2002) Responsive systems for active vibration control. NATO Science Series, Kluwer
3. Varadan VK, Vinoy KJ, Jose KA (2003) RF MEMS and their applications. Wiley-Interscience, Chichester
4. Rebeiz GB (2003) RF MEMS: theory, design, and technology. Wiley-Interscience, Hoboken, NJ, p 350, 2003
5. Pranonsatit S, Lucyszyn S (2005) RF-MEMS activities in Europe. Proceedings of microwave workshops and exhibition 2005:111–122
6. Petersen K (1979) Micromechanical membrane switches on silicon. IBM J Res Dev 23:376–385
7. Larson LE, Hackett RH, Melendes MA, Lohr RF (1991) Micromachined microwave actuator (MIMAC) technology – a new tuning approach for microwave integrated circuits. In: Proceedings of IEEE microwave and millimeter-wave monolithic circuits symposium, 1991, Boston, MA, pp 27–30
8. Kogut L, Komvopoulos K (2004) The role of surface topography in MEMS switches and relays. In: Proceedings of 2004 ASME/STLE international joint tribology conference, 2004, Long Beach, CA, pp 1–5
9. Alper SE, Silay KM, Akin T (2006) A low-cost rate-grade nickel microgyroscope. Sensor Actuat A Phys 132(1):171–181
10. Yi F, Peng L, Zhang J, Han Y (2001) A new process to fabricate the electromagnetic stepping micromotor using LIGA process and surface sacrificial layer technology. Microsyst Technol 7(3):103–106

11. Ledworuski R, Lehr H, Niederfeld G, Walter S, Ruhl F, Tschepe J (2002) A new ultrasonic catheter system with LIGA geared micromotor. Microsyst Technol 9:133–136
12. Gad-el-Hak M (2006) MEMS handbook. CRC Press, Boca Raton, FL
13. Zienkiewicz OC, Taylor RL (2000) Finite element method – the basis. Butterworth-Heinemann, Oxford
14. Ostasevicius V, Gaidys R, Dauksevicius R (2009) Numerical analysis of dynamic effects of a nonlinear vibro-impact process for enhancing the reliability of contact-type MEMS devices. Sensors 9(12):10201–10216
15. Ostasevicius V, Benevicius V, Dauksevicius R, Gaidys R (2009) Theoretical analysis of dynamics of MEMS accelerometer under kinematic excitation. In: Proceedings of Mechanika 2009: 14th international conference, 2–3 Apr 2009, Kaunas, Lithuania, pp 284–290
16. Bagdonas V, Ostasevicius V (2009) Case study of microelectrostatic motor. In: Solid State Phenomena: Mechatronics Systems and Materials III: selected, peer reviewed papers from the 4th international conference: mechatronics systems and materials (MSM 2008), Bialystok, Poland, 14–17 July 2008, pp 113–118
17. Benevicius V, Ostasevicius V, Rimsa G (2009) Sensor nodes in wireless body networks. In: Proceedings of information technologies' 2009: 15th international conference on information and software technologies (IT 2009), Kaunas, Lithuania, 23–24 Apr 2009, pp 125–130
18. Dauksevicius R, Ostasevicius V, Gaidys R (2008) Research of nonlinear electromechanical and vibro-impact interactions in electrostatically driven microactuator. J Vibroeng 10 (1):90–97
19. Ostasevicius V, Dauksevicius R, Gaidys R (2008) Study of natural frequency shifting in a MEMS actuator due to viscous air damping modeled by nonlinear Reynolds equation. J Vibroeng 10(3):388–396
20. Ostasevicius V, Dauksevicius R, Budnikas K (2008) Investigation of cantilever structure used for piezoelectric microgenerators. In: Proceedings of Mechanika 2008: 13th international conference, Apr 3–4 2008, Kaunas, Lithuania, pp 377–381
21. Bagdonas V, Ostasevicius V (2007) Design considerations of a microelectrostatic motor. J Vibroeng 9(4):55–59
22. Bagdonas V, Ostasevicius V (2007) Mathematical modeling and experimental analysis of electrostatic micromotor. In: proceedings of Mechanika 2007: 12th international conference, Kaunas, Lithuania, 5 Apr 2007, pp 20–23
23. Ostasevicius V, Dauksevicius R, Gaidys R, Palevicius A (2007) Numerical analysis of fluid-structure interaction effects on vibrations of cantilever microstructure. J Sound Vib 308 (3–5):660–673
24. Tamulevicius T, Tamulevicius S, Andrulevicius M, Janusas G, Ostasevicius V, Palevicius A (2007) Optical characterization of microstructures of high aspect ratio. In: Proceedings of SPIE: metrology, inspection, and process control for microlithography XXI, vol 6518, 26 Feb–1 Mar 2007, San Jose, CA, pp1–9
25. Ostasevicius V, Bagdonas V, Tamulevicius S, Grigaliunas V (2006) Analysis of a microelectrostatic motor. In Solid State Phenomena: mechatronic systems and materials., pp 185–189
26. Saunoriene L, Ragulskis K, Palevicius A, Ostasevicius V, Janusas G (2006) Hybrid numerical-experimental moire technique for analysis of microstructures. In: Proceedings of SPIE: 7th international conference on vibration measurements by laser techniques: advances and applications, vol 6345, pp 1–8
27. Tamulevicius S, Guobiene A, Janusas G, Palevicius A, Ostasevicius V, Andrulevicius M (2006) Optical characterization of diffractive optical elements replicated in polymers. J Microlith Microfab Microsys 5(1):1–6
28. Dauksevicius R, Ostasevicius V, Tamulevicius S, Bubulis A, Gaidys R (2006) Investigation of electrostatic cantilever-type micromechanical actuator. In: Solid State Phenomena: Mechatronics Systems and Materials: a collection of papers from the 1st international conference (MSM 2005), vol 113, Vilnius, Lithuania, 20–23 Oct 2005, pp 179–184

29. Ostasevicius V, Dauksevicius R, Tamulevicius S, Bubulis A, Grigaliunas V, Palevicius A (2005) Design, fabrication and simulation of cantilever-type electrostatic micromechanical switch. In: Proceedings of SPIE: smart structures and materials 2005: Smart Electronics, MEMS, BioMEMS, and Nanotechnology, vol 5763, 7–10 Mar 2005, San Diego, CA, pp 436–445
30. Ostasevicius V, Tamulevicius S, Palevicius A, Ragulskis MK, Ramutis P, Grigaliunas V (2005) Hybrid numerical-experimental approach for investigation of dynamics of microcantilever relay system. Opt Lasers Eng 43(1):63–73
31. Ostasevicius V, Tamulevicius S, Palevicius A, Ragulskis MK, Grigaliunas V, Minialga V (2005) Synergy of contact and noncontact techniques for design and characterization of vibrating MOEMS elements. J Microlith Microfab Microsys 4(4):1–9
32. Ragulskis L, Ragulskis MK, Palevicius A, Ostasevicius V, Palevicius R (2005) Shearographic technique for NDE analysis of high frequency bending vibrations of microstructures. In: Proceedings of SPIE: testing, reliability, and application of micro- and nano-material systems III: 3rd conference on testing, reliability, and application of micro and nano-material systems, vol 5766, 08–10 Mar 2005, San Diego, CA, pp 118–125
33. Ostasevicius V, Ragulskis MK, Palevicius A, Kravcenkiene V, Janusas G (2005) Applicability of holographic technique for analysis of non-linear dynamics of MEMS switch. In: Proceedings of SPIE: smart structures and materials 2005: smart electronics, MEMS, BioMEMS, and Nanotechnology, vol 5763, 7–10 Mar 2005, San Diego, CA, pp 405–413
34. Ostasevicius V, Gitis N, Palevicius A, Ragulskis MK, Tamulevicius S (2005) Hybrid experimental-numerical full-field displacement evaluation for characterization of micro-scale components of mechatronic systems. In: Solid state phenomena: mechatronics systems and materials: a collection of papers from the 1st international conference (MSM 2005), vol 113, Vilnius, Lithuania, 20–23 Oct 2005, pp 73–78
35. Palevicius A, Ragulskis K, Bubulis A, Ostasevicius V, Ragulskis MK (2004) Development and operational optimization of micro spray system. In: Proceedings of SPIE: smart structures and materials 2004: smart structures and integrated systems, vol 5390, 15–18 Mar 2004, San Diego, CA, pp 429–438
36. Ostasevicius V, Palevicius A, Daugela A, Ragulskis MK, Palevicius R (2004) Holographic Imaging technique for characterization of MEMS switch dynamics. In: Proceedings of SPIE: smart structures and materials 2004: smart electronics, MEMS, BioMEMS, and nanotechnology, vol 5389, 15–18 Mar 2004, San Diego, CA, pp 73–84

Chapter 2
Modeling and Simulation of Contact-Type Electrostatic Microactuator

Abstract This chapter presents 3-D FE modeling and simulation of dynamics of microcantilever operating in ambient air near fixed surface. The phenomenon of squeeze-film damping is further analyzed numerically. Frequency response and transient analyses are carried out in order to determine influence of squeeze-film damping on free and forced vibrations of the microcantilever under different ambient air and vibration excitation conditions. Subsequently numerical analysis of the 3-D microcantilever under the effect of electrostatic field is provided. Static and dynamic simulations are performed in order to study important operational characteristics. Finally, the chapter is concluded with FE modeling of the microcantilever with incorporated adhesive-repulsive contact model, which uses a "classical" linear elastic link element combined with the van der Waals force-based term that accounts for the influence of dominant intermolecular interactions in the contact zone. This model is then used in conjunction with squeeze-film damping formulation in order to predict behavior of contact bouncing under different air damping and contact conditions.

2.1 Three-Dimensional Finite Element Model of a Microcantilever

A microstructure having cantilever or fixed–fixed configuration is a basic structural element of most microelectromechanical actuators and sensors such as microswitches, capacitive pressure sensors, microaccelerometers, filters, resonators and many others. Structural element of the developed electrostatically-actuated microswitch, presented in Chapter 5, is a bi-metal, complex-shaped microstructure with non-ideal cantilever-type fixing conditions. During modeling it is approximated by an ideal homogeneous microcantilever having material properties of bulk nickel. This idealization is adopted since actual fixing conditions and material properties of the original bi-metal structure are not known a priori and their experimental

V. Ostasevicius and R. Dauksevicius, *Microsystems Dynamics*,
Intelligent Systems, Control and Automation: Science and Engineering 44,
DOI 10.1007/978-90-481-9701-9_2,

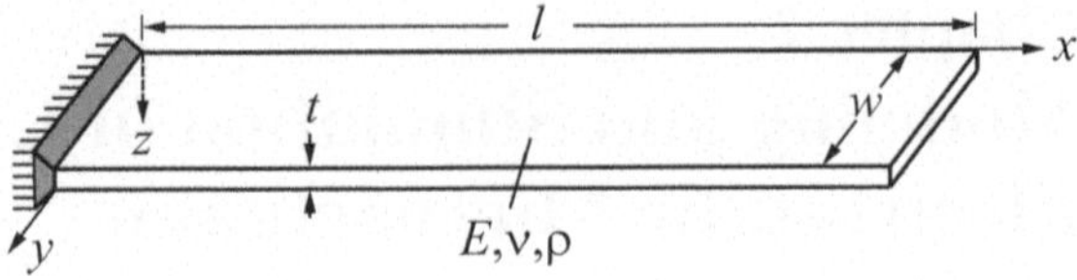

Fig. 2.1 Schematics of a microcantilever that is used for numerical modeling

Table 2.1 Typical values of microcantilever parameters used for numerical analysis

Description and symbol	Value	Unit
Length l	67÷117	μm
Width w	30	μm
Thickness t	2.0	μm
Young's modulus E	207 [1]	GPa
Density ρ	8902 [1]	Kg/m^3
Poisson's ratio ν	0.31 [1]	-

characterization would require development of complicated microscale-adapted experimental setup and testing methodology.

Schematic drawing of a basic modeled microcantilever with typical parameter values is provided in Fig. 2.1 and Table 2.1 respectively. Firstly, mechanical model of the microcantilever was created in the commercial finite element (FE) modeling software Comsol Multiphysics (further – COMSOL). In the FE formulation microcantilever dynamics is described by the following equation of motion presented in a general matrix form:

$$[M]\{\ddot{U}\} + [C]\{\dot{U}\} + [K]\{U\} = \{Q(t, U, \dot{U})\}, \tag{2.1}$$

where $[M]$, $[C]$, $[K]$ are mass, damping and stiffness matrices respectively, $\{U\}$, $\{\dot{U}\}$, $\{\ddot{U}\}$ – displacement, velocity and acceleration vectors respectively, $Q\{t, U, \dot{U})\}$ – vector representing the sum of external forces acting on the microcantilever. This load vector represents one of the following forces or a sum of these: (a) air pressure forces generated by fluidic–structural interaction between the microcantilever and the surrounding air (see Section 2.2), (b) electrostatic forces generated by electrostatic–structural interaction between the microcantilever and the applied electrostatic field (see Section 2.3), (c) contact forces generated by vibro-impact interaction between the microcantilever and a stationary structure (see Section 2.4).

For the coupled-field analysis of the fluidic-structural interaction, the mechanical model of the microcantilever was coupled in COMSOL with one of the three different forms of Reynolds equation as detailed in Section 2.2. Nonlinear and compressible forms of this equation originally were not present in the software therefore they were inserted into it thereby expanding the capabilities of the software. For the coupled-field analysis of the electrostatic-structural interaction, the mechanical model of the microcantilever was integrated with additional built-in COMSOL modules. Afterwards, this model was further developed in order to make

Table 2.2 Simulated natural frequencies of transverse vibrations of microcantilevers

Length of microcantilever l and order of natural frequency	Natural frequency of transverse vibrations f (kHz)
$l = 67$ μm:	
f_1	355.9
f_2	2,211.8
f_3	6,175.6
$l = 87$ μm:	
f_1	210.3
f_2	1,311.3
f_3	3,671.8
$l = 117$ μm:	
f_1	116.1
f_2	727.2
f_3	2,048.2

it suitable for simulation of actual electrostatic actuation conditions as detailed in Section 2.3. For the analysis of the vibro-impact interaction, the mechanical model of the microcantilever was integrated with a specially-derived adhesive-repulsive contact model as detailed in Section 2.4.

The developed mechanical model was first subjected to numerical modal analysis for prediction of natural frequencies, which are listed in Table 2.2.

2.2 Finite Element Modeling and Simulation of Fluidic–Structural Interaction

2.2.1 Model Formulation

Fluidic–structural interaction in a microswitch (or in any other MEMS actuator or sensor, which operation is based on vertical movement of active microstructure in ambient gas environment in close proximity to a second stationary structure) manifests as squeeze-film damping, which is a counter-reactive pressure force generated by squeezed gas film in the gap. Figure 2.2 shows a schematic drawing of a model that is used for numerical analysis of influence of squeeze-film damping on vibrations of microcantilever. The model consists of a flexible 3-D microcantilever, which is fixed at $x = 0$ and is suspended by distance h_0 over a stationary structure (representing substrate).

Parameters listed in Table 2.3 are typically used in this chapter unless specified otherwise. It is assumed that both structures are surrounded by the air. Therefore the gap in-between is filled by the air and forms an air-film, which is characterized by the following parameters that are specified in COMSOL before the problem solution in a

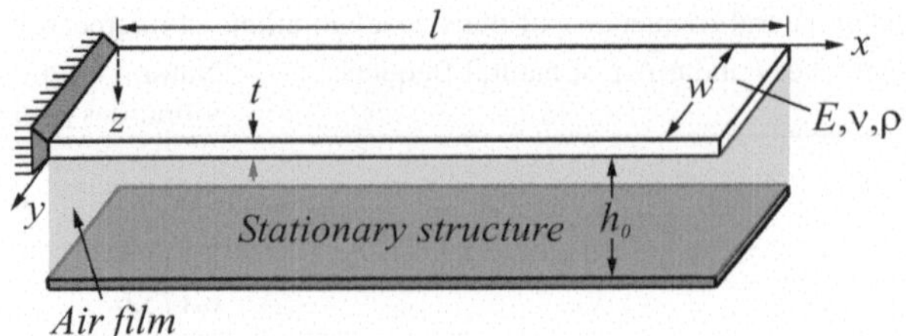

Fig. 2.2 Schematic representation of a modeled fluidic–structural system consisting of a microcantilever under the effect of forces generated by squeezed air film

Table 2.3 Typical parameter values used for numerical analysis of influence of squeeze-film damping on vibrations of microcantilever

Description and symbol	Value	Unit
Length, l	117	μm
Width, w	30	μm
Thickness, t	2.0	μm
Young's modulus, E	207	GPa
Density, ρ	8902	Kg/m^3
Poisson's ratio, ν	0.31	-
Dynamic viscosity of air μ at STP [2]	18.45×10^{-6}	Pa·s
Mean free path of air particles at atmospheric pressure L_0 [3]	65	nm
Air-film thickness h_0	$0.5 \div 4$	μm

pre-processing stage: h_0 – nominal air-film thickness, μ – dynamic viscosity of air at standard pressure and temperature STP (101 kPa, 25°C), L_0 – mean free path of air particles at atmospheric pressure and p_0 – working pressure (i.e. initial ambient pressure). The microcantilever is initially at rest in its undeformed configuration.

Three versions of Reynolds equation are used for the modeling of pressure variation in the gap due to squeezed air-film effect:

Nonlinear Reynolds equation (NRE):

$$\frac{\partial}{\partial x}\left(\frac{Ph^3}{\mu_{eff}}\frac{\partial P}{\partial x}\right)+\frac{\partial}{\partial y}\left(\frac{Ph^3}{\mu_{eff}}\frac{\partial P}{\partial y}\right)=12\left(h\frac{\partial P}{\partial t}+P\frac{\partial h}{\partial t}\right). \tag{2.2}$$

Linearized Reynolds equation (LRE):

$$\frac{p_0 h_0^2}{12\mu_{eff}}\left[\frac{\partial^2}{\partial x^2}\left(\frac{\delta P}{p_0}\right)+\frac{\partial^2}{\partial y^2}\left(\frac{\delta P}{p_0}\right)\right]=\frac{\partial}{\partial t}\left(\frac{\delta h}{h_0}\right)+\frac{\partial}{\partial t}\left(\frac{\delta P}{p_0}\right). \tag{2.3}$$

Linearized incompressible Reynolds equation (LIRE):

$$\frac{h_0^3}{12\mu_{eff}}\left[\frac{\partial^2}{\partial x^2}\left(\frac{\delta P}{p_0}\right)+\frac{\partial^2}{\partial y^2}\left(\frac{\delta P}{p_0}\right)\right]=\frac{\partial h}{\partial t}, \tag{2.4}$$

where h is the variable distance between the microcantilever and the ground, $P = p_0 + \Delta p$, (P – total pressure in the gap, Δp – an additional film pressure due to the squeezed air-film effect), μ_{eff} is the effective viscosity of the air in the gap, which accounts for the rarefied gas effects due to low working pressure by being dependent on the Knudsen number K_n. The model of Veijola [2, 4, 5] is used for μ_{eff}. It is valid over a wide range of Knudsen numbers ($0 \le K_n \le 880$) with 5% accuracy:

$$\mu_{eff} = \frac{\mu}{1 + 9.638 {K_n}^{1.159}}, \tag{2.5}$$

$$K_n = \frac{L_0 P_{atm}}{p_0 h_0}. \tag{2.6}$$

Air can freely move into and out of the gap. Due to transient changes in the gap size, an additional time-dependent pressure component Δp appears in the air gap which depends on the gap size and its deformation velocity as well as on the properties of the air and the structures. Because the system's surroundings are in equilibrium, the only force component that affects the moving microcantilever results from the additional film pressure Δp. At open edges of the microcantilever the pressure is equal to atmospheric, i.e. the corresponding end condition is $\Delta p = 0$. This is an adequate assumption since the aspect ratio (ratio between lateral dimensions and thickness of the air-film) of the considered microsystem is large and therefore variation of Δp vanishes at the edges of the gap.

Modeling and simulation of the considered microsystem is performed by means of finite elements method within COMSOL package. Due to symmetry only half of the microcantilever is modeled as it is indicated by special boundary condition symbols in Fig. 2.3 (right). The symmetry is not employed only in the modal analysis. In order to associate the air-film with the microcantilever for the subsequent numerical analysis of coupled fluidic-structural problem, above-presented versions of Reynolds

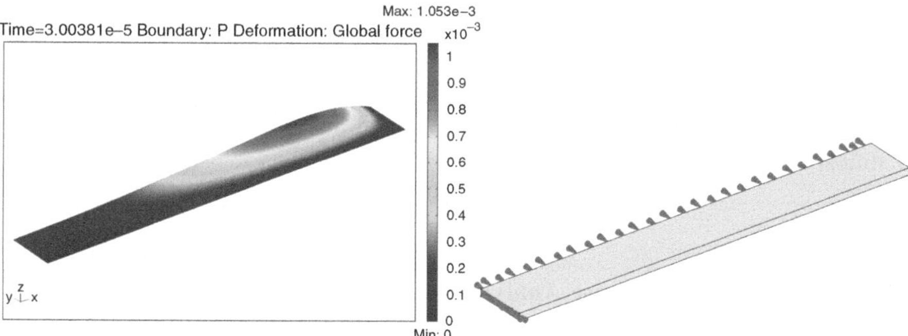

Fig. 2.3 (*Left*) pressure distribution in the gap with air at atmospheric pressure after 30 μs of transient simulation of free vibrations of microcantilever. (*Right*) COMSOL finite element half-symmetry model of microcantilever. Special symbols located next to the structure indicate boundary conditions (clamping of one end and constraint of symmetry plane)

equation are transformed into weak formulation, then inserted into the FE model and "coupled" to a lower surface (boundary) of the microcantilever. It is possible to insert and use the required equations or expressions in a COMSOL model due to the unique feature of this FEM software that is referred to as "equation-based modeling", which enables the user to input any number of equations via graphical user interface (GUI). And then the package uses a special equation interpreter to automatically translate the expression into a finite element code. COMSOL solves entered Reynolds equations in order to determine pressure distribution in the gap and then this pressure acts as a boundary load on the microcantilever. In total, three separate finite element models have been developed: in each model the microcantilever is coupled with different version of Reynolds equation – NRE, LRE and LIRE.

2.2.2 *Numerical Analysis of Influence of Squeeze-Film Damping on Vibrations of Microcantilever*

2.2.2.1 Modal Analysis

Numerical analysis of squeeze-film damping begins with modal analysis, which purpose is to determine the distribution of air pressure forces in the gap when microcantilever is vibrating in its flexural and torsional resonant modes. For these simulations LIRE-based model was used and the working pressure p_0 in the gap was equal to atmospheric value of $P_{atm} = 101325$ Pa. COMSOL eigenfrequency solver was used in order to perform this modal analysis. The results, presented in Fig. 2.4, consist of several natural frequencies of the microcantilever, corresponding structural mode shapes and the associated pressure mode shapes. When examining these results we may notice the obvious coupling between structural displacements of the microcantilever and pressure distribution in the gap. For example, in the second flexural mode, the upward flexing of middle part of microcantilever corresponds to a concave pressure profile in the respective region of pressure distribution plot, which indicates the reduction of pressure in this part of the gap (i.e. decompression effect). And, in contrast, the downward flexing of free end of the microcantilever corresponds to a convex pressure profile – zone of increased pressure with respect to atmospheric (i.e. compression effect).

2.2.2.2 Frequency Response Analysis

Frequency response analysis was performed with dual purpose: (1) to verify developed COMSOL squeeze-film damping model by determining amplitude–frequency characteristics, extracting the associated quality factors and comparing their values with results obtained from available approximate analytical expressions,

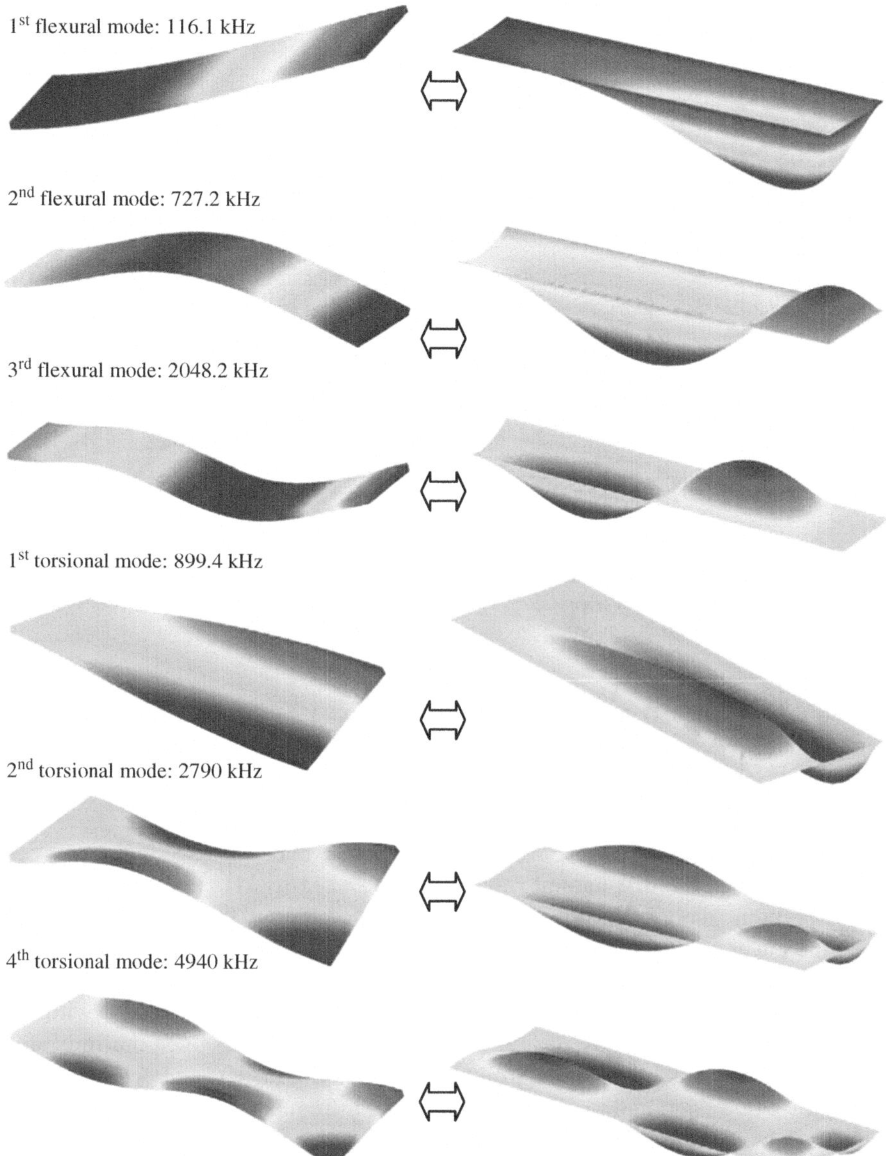

Fig. 2.4 (*Left*) simulated flexural and torsional vibration mode shapes of microcantilever. (*Right*) simulated pressure mode shapes – 3-D contour plots illustrating distribution of air pressure forces in the gap for the corresponding structural mode shapes

(2) to study the phenomenon of squeeze-film damping by investigating simulated amplitude–frequency characteristics obtained with different system parameters.

Thus, frequency response analysis was used in order to determine transient response of the microcantilever throughout the range of specified harmonic

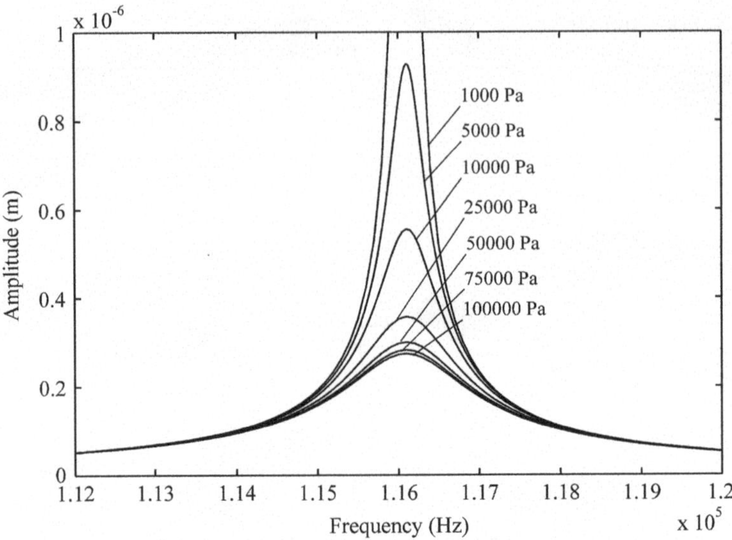

Fig. 2.5 Amplitude–frequency characteristics of arbitrary point of microcantilever in the vicinity of its first natural frequency of transverse vibrations and in the presence of squeeze-film damping at different levels of working pressure p_0 ($h_0 = 4$ μm, $l = 117$ μm, $\sigma = 0.08$ at $p_0 = P_{atm}$)

excitation frequencies at different ambient pressures p_0. For this purpose a harmonic load F_e was applied on the lower edge of the free end of the microcantilever:

$$F_e = F_a \sin(2\pi f_e t). \tag{2.7}$$

Range of excitation frequencies of $f_e = 112 \div 120$ kHz was used in the analysis because it is near the first natural frequency of $f_1 = 116.1$ kHz that was estimated during previous modal analysis. Force amplitude of $F_a = 0.001$ N/m was applied along the lower edge, which is 30 μm wide. LRE-based model was used in this type of numerical analysis. During this and further simulations of squeeze-film damping the structural damping was equal to zero. Parametric solver, associated with the frequency response analysis, was applied in order to sweep over the indicated frequency range and thereby determine amplitude–frequency characteristic presented in Fig. 2.5. Quality factors Q for each frequency response plot were calculated by using the following expression:

$$Q \approx \frac{1}{2\zeta_r} = \frac{f_r}{\Delta f}, \tag{2.8}$$

where $f_r = f_1$ – first natural frequency of analyzed microcantilever, ζ_r – damping ratio at resonance frequency f_r, $\Delta f = f_{(2)} - f_{(1)}$ is the bandwidth, which represents the distance between two points on frequency axis where the amplitude is equal to $1/\sqrt{2} \approx 0.707$ of the maximum amplitude value.

Table 2.4 Comparison of quality factors of microcantilever extracted from results of numerical frequency response analysis with squeeze-film damping by using COMSOL model with values calculated by using approximate analytical expressions for quality factors of microcantilevers

Working pressure p_0 (kPa)	Authors' COMSOL model (3-D)	Approximate analytical expression for microcantilever Q^* [2, 6]	Approximate analytical expression for microcantilever Q^{**} [7]
0.01	1935	2.48×10^5	2.52×10^5
0.1	1659	1.72×10^4	1.75×10^4
1	1161	1239	1260
5	237	232	237
10	116	131	133
25	94	76	77
50	83	61	62
75	77	56	57
100	73	54	55

These extracted quality factors were compared against results obtained with available approximate analytical formulas for Q^* [2, 6] and Q^{**} [7] that have been derived for microcantilevers:

$$Q^* = \frac{\sqrt{E\rho}t^2}{\mu_{eff}(wl)^2}h_0^3, \tag{2.9}$$

$$Q^{**} = \frac{\rho h_0^3 t}{\mu_{eff} w^2}\omega_1^2. \tag{2.10}$$

These results are summarized in Table 2.4. It is obvious that the predicted and analytical Q values are in a reasonably good agreement in the pressure range of $p_0 = 1 \div 100$ kPa, which is referred to as "viscous region" because air is treated here as viscous fluid. Meanwhile, in the range of $p_0 = 0.01 \div 0.1$ kPa analytical Q values are considerably higher than obtained from simulation, i.e. in comparison to the COMSOL model, the analytical models underestimate the magnitude of air damping in this rarefied air range, which is referred to as "molecular region" because here damping is caused by independent collisions of noninteracting air molecules with the surface of vibrating structure. However, one should keep in mind that these analytical expressions are approximate: Q^* was derived by using Stokes law, which is applied for viscous flow therefore its accuracy in molecular region is very doubtful, meanwhile Q^{**} was obtained from highly simplified Reynolds equation and its results beyond viscous region are also unreliable. On the other hand, developed FEM model accounts for air rarefaction effects by means of modification of dynamic viscosity μ and is suitable for modeling of air damping effects in microsystems having ambient pressure in the range of $p_0 = 0.01 \div 100$ kPa and air-gaps of $h_0 \geq 1$ μm (i.e. when $0 \leq K_n \leq 880$).

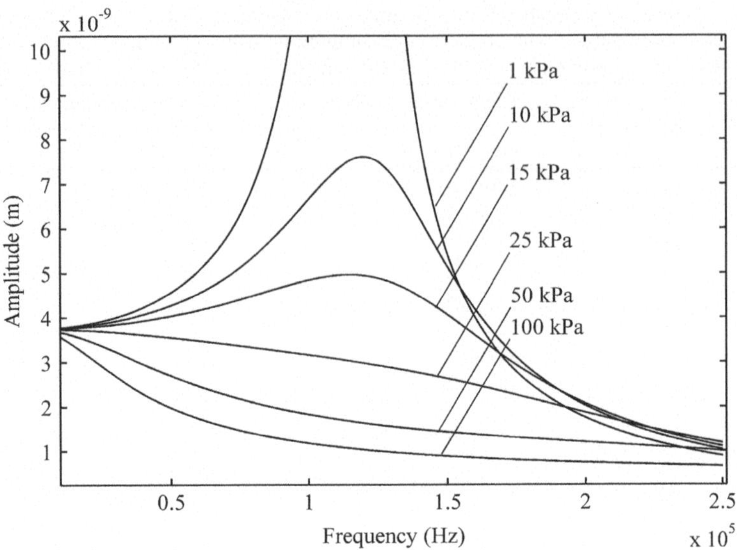

Fig. 2.6 Amplitude–frequency characteristics of arbitrary point of the microcantilever in the vicinity of the first natural frequency and in the presence of squeeze-film damping modeled with LRE ($h_0 = 0.5$ μm, $l = 117$ μm)

Quality factors listed in Table 2.4 indicate that in this particular case the level of air damping is relatively low even at atmospheric pressure. It should be noted that these numerical results were obtained with air-film thickness of $h_0 = 4$ μm. However, microsystems can be designed and fabricated with air gaps as small as 0.5 μm. Therefore one more series of harmonic analyses was performed in the frequency range of f_e=10 ÷ 250 kHz in order to study the influence of the reduced air-film thickness on the frequency response of the microcantilever. The results, illustrated in Fig. 2.6, reveal the existence of significant damping for $h_0 = 0.5$ μm. The resonance peak has already been suppressed for the working pressure of $p_0 = 25$ kPa, which in turn implies that at atmospheric pressure the resonance could be suppressed even at larger values of h_0. These results demonstrate that under the effect of squeeze-film damping the quality factors of the analyzed microsystem may be reduced to as low as $Q \leq 1$.

It is well known that in the case of squeeze-film damping phenomenon pressure force exerted by the air-film undergoing periodic cycles of compression and decompression has two components: one is in phase with the microcantilever velocity (i.e. viscous damping force) and the other is in phase with the displacement (i.e. elastic force component due to compressibility of air). A non-dimensional squeeze number σ is used to characterize the degree of compression of the air in the gap:

$$\sigma = \frac{12\mu_{eff}L^2\omega}{p_0 h_0^2}, \tag{2.11}$$

where L is the characteristic length – the shortest lateral dimension of the microcantilever.

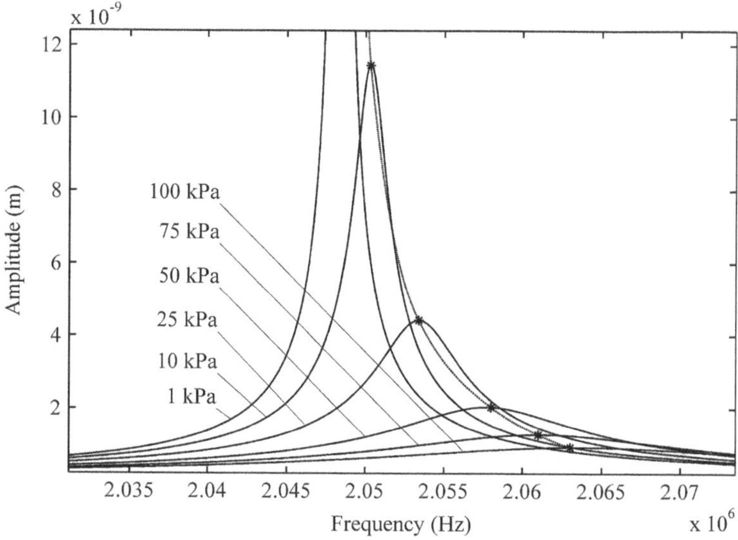

Fig. 2.7 Amplitude–frequency characteristics of arbitrary point of microcantilever in the vicinity of its third natural frequency of transverse vibrations and in the presence of squeeze-film damping at different levels of working pressure p_0 ($h_0 = 1$ μm, $l = 117$ μm, $f_3 = 2{,}0482$ MHz, $\sigma = 16.1$ at $p_0 = P_{atm}$). Dashed line indicates the trend of natural frequency shifting

At relatively low oscillation frequencies ω or relatively large air gaps h_0 (i.e. low squeeze number – roughly$\sigma \leq 3$ [2]), the viscous damping force dominates because the air can escape out of the gap without being compressed. While, at relatively high ω or relatively small h_0 (i.e. high σ), elastic forces increase because of the air-film compression effect. In such conditions, the film becomes virtually non-dissipative, exhibiting a nonlinear spring action. In practice, the squeezed air-film represents a combination of viscous damping and elastic forces [8–12].

The main effect of air compression at high values of σ is the stiffening of the microsystem, which consequently increases its natural frequency. Results from frequency response analysis performed in the vicinity of the first natural frequency of transverse vibrations (Fig. 2.5) did not reveal natural frequency shifts since squeeze number for this case is very low, just $\sigma = 0.08$ at $p_0 = P_{atm}$. Therefore, another sequence of frequency response analyses was carried out in order to study the phenomenon of natural frequency shifting.

The condition of larger σ was achieved by increasing harmonic excitation frequencies and therefore the simulation was performed in the vicinity of third natural frequency of transverse vibrations of typical microcantilever ($f_3 = 2.0482$ MHz). The squeeze number for this case was equal to $\sigma = 16.1$ at $p_0 = P_{atm}$. Obtained amplitude–frequency characteristics, provided in Fig. 2.7, clearly demonstrate that in this case air undergoes compression and this in turn raises the natural frequency of the microcantilever as the pressure increases from 1 to 100 kPa. Figure 2.8 illustrates the change of relative frequency shift as a function of p_0. Simulation results reveal that

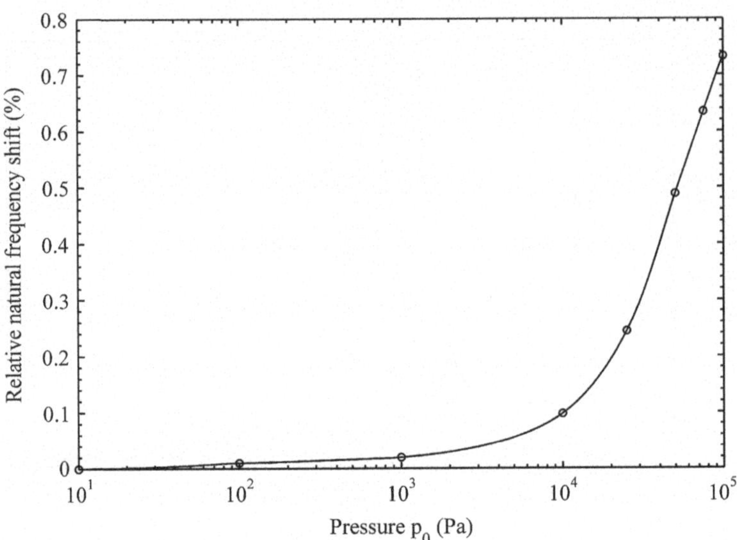

Fig. 2.8 Relative shift of the third natural frequency of transverse vibrations of the microcantilever presented as a function of working pressure p_0 ($h_0 = 1$ μm, $l = 117$ μm, $f_3 = 2.0482$ MHz, $\sigma = 16.1$ at $p_0 = P_{atm}$)

the third natural frequency of transverse vibrations in air at $p_0 = P_{atm}$ is by 0.73% higher than the value obtained under vacuum conditions ($p_0 = 0$). It should be pointed out that no significant pressure dependence was observed below 100 Pa and this result is in agreement with experimental results and observations found in open literature [13].

2.2.2.3 Time-Dependent Analysis

Case of free vibrations. In order to determine influence of squeeze-film damping effects on free vibrations of the microcantilever, the latter is set to oscillate freely by displacing upwards its lower edge at the free end by a certain distance z_0 (static analysis) and then releasing (transient analysis). When the value of z_0 is such that $h_0/z_0 \approx 1$, then the resulting free vibrations are considered to be large-amplitude (with respect to air-film thickness h_0) and when $h_0/z_0 \approx 100$ – small-amplitude. The above-described procedure was used to perform a numerical analysis, which purpose was to determine influence of variation of working pressure p_0 and air-film thickness h_0 on magnitude of squeeze-film damping. Numerical results illustrated in Fig. 2.9 were obtained with LIRE-based model and using large-amplitude

Fig. 2.9 (continued) *Middle*: at the fixed working pressure of $p_0 = 10^5$ Pa but for different values of nominal gap thickness h_0. Curves: 1–1, 2–2, 3–3 μm. *Bottom*: typical variation of additional film pressure Δp at arbitrary mid-point located near the free end of microcantilever, which undergoes decaying free oscillations. Each curve represents a case of different working pressure p_0

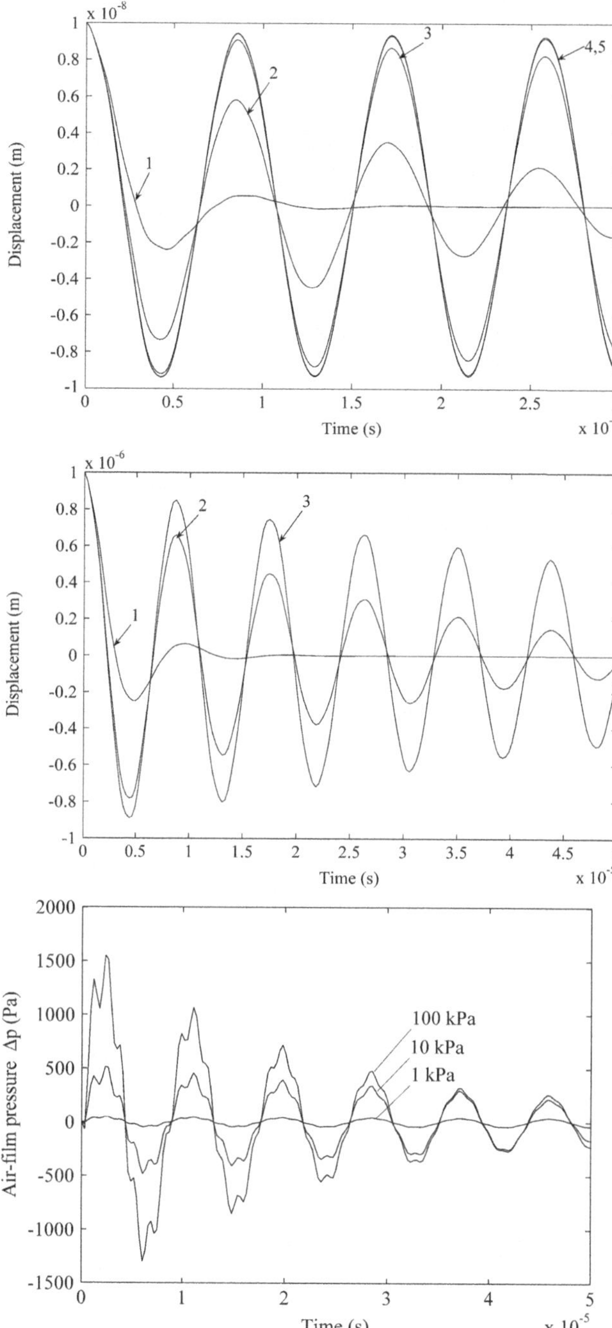

Fig. 2.9 Simulated results from model with LIRE showing free vibration response curves of the arbitrary point at the end of a microcantilever after it has been released from its initial deflected position $z_0 = 1$ μm upwards. *Top:* for the different levels of working pressure p_0 in the gap with thickness $h_0 = 1$ μm. Curves: 1–10^5, 2–10^4, 3–10^3, 4–10^2, 5–10 Pa.

oscillations. These curves reveal that both parameters have a significant effect on magnitude of air damping: vibration decay and, consequently, squeeze-film damping intensifies with increase in working pressure p_0 and decrease in gap thickness h_0. As an example, Fig. 2.3 (left) visualizes pressure distribution in the gap with air at $p_0 = P_{atm}$ after 30 μs of transient simulation. The convex profile of pressure distribution indicates areas of increased pressure in the gap with respect to the atmospheric value. In these areas squeeze-film phenomenon is pronounced the most and the pressure cycle changes from compression to decompression while the microcantilever is undergoing decaying periodic motion. This figure indicates that near the edges of the microcantilever pressure is equal to atmospheric, which corresponds to the imposed boundary conditions for pressure. In addition, Fig. 2.9 illustrates variation of additional air-film pressure Δp at arbitrary point (located in the middle of microcantilever towards its free end) when the microcantilever is undergoing decaying periodic motion.

Amplitude effects. NRE-based model was used to perform transient simulations of free vibrations of microcantilever in the case of large-amplitude motion for different levels of p_0 in the gap. The results of this numerical analysis are provided in Fig. 2.10 (top). By observing the time-coordinate of amplitude peaks of curves corresponding to different levels of p_0 (t_1, t_2, t_3, t_4, t_5) it is possible to notice that the time-coordinate of the peak decreases (from t_5 to t_1) with increase in p_0. In the case of model with LIRE, the peak position on x-axis is constant as can be noticed in Fig. 2.9. Figure 2.10 (bottom) illustrates the variation of time-coordinate of the peaks with respect to working pressure p_0 represented on logarithmic scale. It is obvious that the most pronounced reduction is from t_3 to t_1 as p_0 changes from 1 to 100 kPa. Total relative decrease from t_5 to t_1 is 16.3%. This effect also manifests in the case of small-amplitude free vibrations but in this case it is not so significant. Reduction of time-coordinate of amplitude peak with increasing pressure indicates that the frequency of natural vibrations increases and this in turn implies that in this case air undergoes compression leading to natural frequency shift.

Model comparison. Multiple transient simulations were carried out with different squeeze-film damping models in order to compare the magnitude of air damping that each model generates under different working pressure in the gap in the case of free vibrations. Damping ratios were subsequently extracted from the obtained decaying vibration curves by using the following equation [14]:

$$\zeta = \frac{\Delta}{\sqrt{4\pi^2 + \Delta^2}}, \tag{2.12}$$

where ζ – damping ratio, Δ – logarithmic decrement.

The extracted damping ratios were recalculated to quality factors. Results of this analysis, summarized in Fig. 2.11 (top-left), reveal that for the case of free vibrations, amount of induced squeeze-film damping estimated with different models is very similar, only the difference between NRE and LIRE as well as between LRE and LIRE is more pronounced. It is obvious from the plot of Fig. 2.11 (top-left) that LIRE-based model underestimates air damping in comparison to LRE- and NRE-

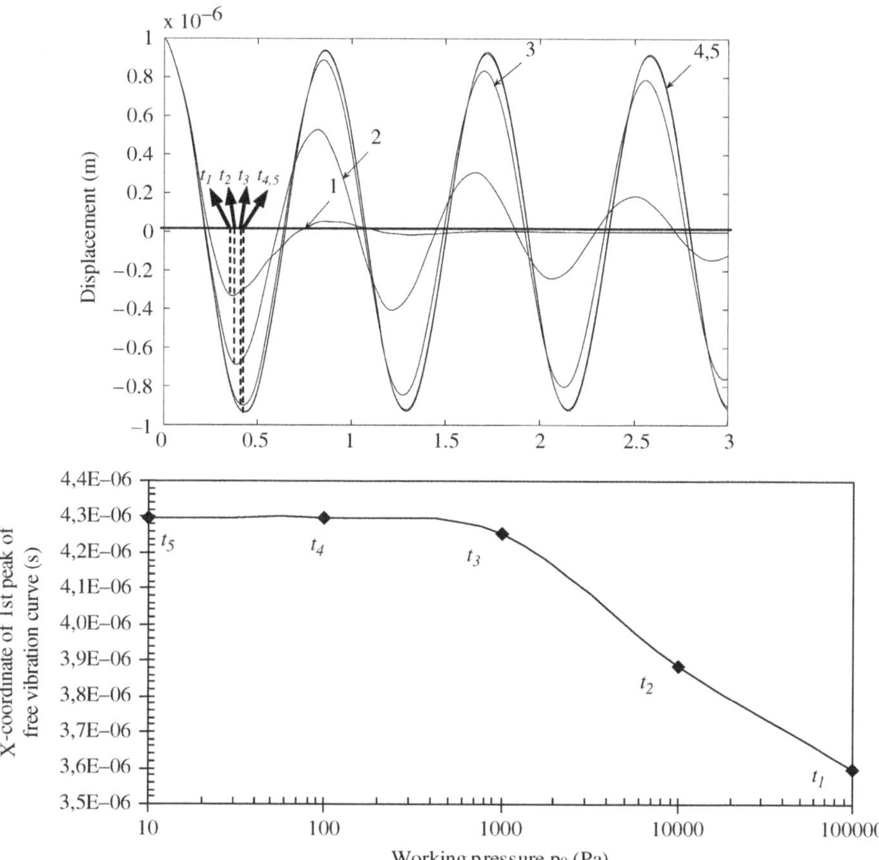

Fig. 2.10 *Top*: results from model with NRE showing free vibration response curves of the arbitrary point at the end of the microcantilever after it was released from its initial deflected position $z_0 = 1$ μm for different p_0 in the gap of $h_0 = 1$ μm. Curves: 1–10^5, 2–10^4, 3–10^3, 4–10^2, 5–10 Pa. *Bottom*: variation of time-coordinate of peaks of free vibration response curves corresponding to different values of working pressure p_0 as indicated in (**a**) with t_1, t_2, t_3, t_4 and t_5

based models. Simulation results indicate that the largest difference in quality factors between NRE and LIRE is obtained for working pressures $p_0 = 100$ Pa and $p_0 = 1000$ Pa: Q values from LIRE-based model are higher than those from NRE-based model by 20% and 14% respectively. Results from NRE- and LRE-based models are nearly coincidental with only 0.7% difference. For verification purposes, the figure also contains analytical values of quality factors calculated by using expression for Q^* in Eq. (2.9). Similarly to the results obtained from frequency response analysis, simulated and analytical values of quality factors match each other reasonably well in the pressure range of $p_0 = 1 \div 100$ kPa and constitute the difference of one or two orders of magnitude when $p_0 = 0.01 \div 0.1$ kPa. One may notice from the graphical results that the difference between these values increases as the pressure decreases.

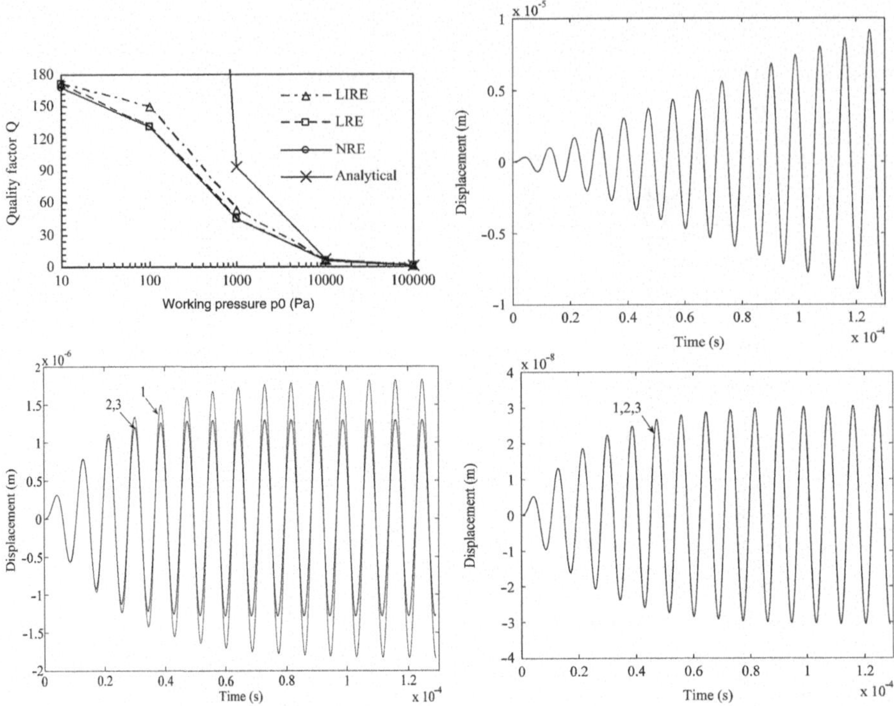

Fig. 2.11 *Top-left*: comparison of analytical and simulated quality factors Q as a function of p_0. Simulated Q extracted from free vibration response curves ($z_0 = 1$ μm, $h_0 = 1$ μm). *Top-right*: simulated response of arbitrary end point of undamped microcantilever excited with sinusoidal force with $f_e = f_1$. *Bottom*: time response of arbitrary end point of microcantilever excited with sinusoidal force with $f_e = f_1$ having magnitude that produces large-amplitude (*left*) and small-amplitude ($h_0/z_0 \approx 70$) (*right*) vibrations in the presence of squeeze-film damping modeled with LIRE (1), LRE (2) and NRE (3) ($h_0 = 2$ μm, $p_0 = 10^5$ Pa)

Case of forced vibrations. For the evaluation of squeeze-film damping effects in the case of forced vibrations, lower edge of the free end of the microcantilever was excited with sinusoidal force $F_a\sin(2\pi f_e t)$. During the numerical analysis effects of small- and large-amplitude motions (achieved by specifying appropriate force amplitude F_a) as well as influence of different excitation frequency f_e on squeeze-film damping were predicted. During the simulations working pressure in the gap was constant and set to $p_0 = 10^5$ Pa.

Time response of arbitrary end point of microcantilever was investigated with all three damping models in the case when the microcantilever was excited with force, which frequency is equal to fundamental frequency of the microcantilever ($f_e = f_1 = 116.1$ kHz). When sinusoidal excitation of this frequency is applied to the microcantilever without Reynolds equation coupled to it (i.e. without squeeze-film damping), then we obtain amplitude increase without bound producing resonance as illustrated in Fig. 2.11 (top-right). Figure 2.11 (bottom-right) shows time response of arbitrary end point of microcantilever when it is excited with sinusoidal force of the magnitude that produces small-amplitude vibrations ($h_0/z_0 \approx 70$) and which frequency is equal to fundamental one in the presence of squeeze-film damping modeled with LIRE, LRE and NRE. It is obvious that curves obtained with different models coincide very closely. However different results are observed when F_a is increased so as to generate large-amplitude vibrations (Fig. 2.11 (bottom-left)). For this case it was determined that response amplitude for the model with LIRE is about 35% larger than with LRE and NRE. This leads to the conclusion that model with LIRE underestimates squeeze-film damping of microcantilever in comparison to models with LRE and NRE, which results again are nearly coincidental.

A series of simulations were performed in order to compare transient responses of velocity and pressure at arbitrary midpoint at the free end of microcantilever excited with sinusoidal force of the magnitude that produces large-amplitude vibrations in the presence of squeeze-film damping modeled with LIRE (Fig. 2.12 (top))

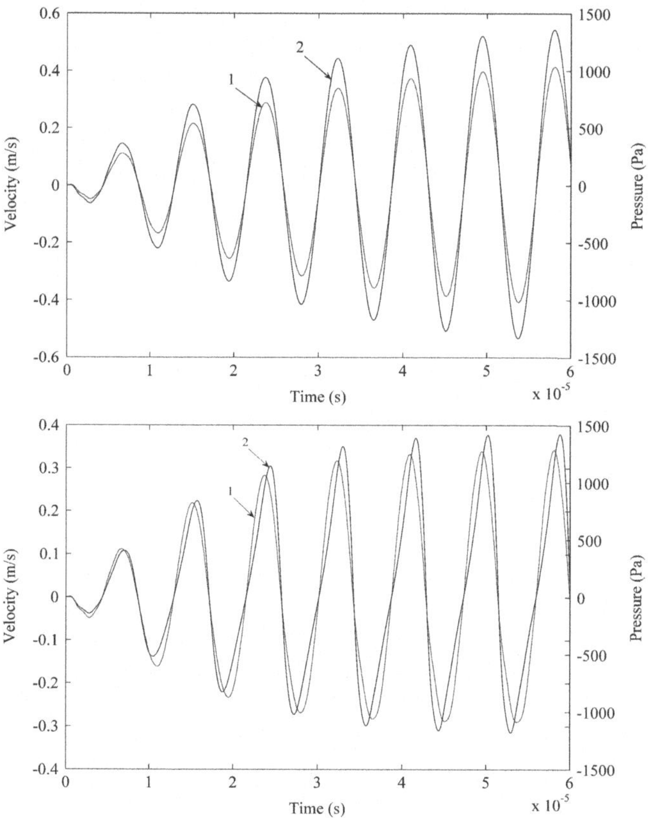

Fig. 2.12 Comparison of time responses of velocity (1) and pressure (2) at arbitrary midpoint of microcantilever excited with sinusoidal force of the magnitude that produces large-amplitude vibrations and which frequency is equal to fundamental frequency of microcantilever in the presence of squeeze-film damping modeled with: LIRE (*top*), NRE (*bottom*) ($h_0 = 2$ μm, $p_0 = 10^5$ Pa)

and NRE (Fig. 2.12 (bottom)). These results reveal obvious differences between the models – Fig. 2.12 (bottom) indicates a nonlinear pressure response for large-amplitude motion.

In order to further study air compressibility effects due to squeeze-film damping phenomenon, another sequence of transient simulations was carried out. Variation of frequency of sinusoidal excitation f_e in the course of numerical analysis allowed to determine the influence of excitation frequency on character of squeeze-film damping as well as emphasize the differences between the models. Fig. 2.13 (top) and Fig. 2.13 (middle) demonstrate transient responses of velocity (curve 1) and pressure (curve 2) at arbitrary midpoint at the free end of microcantilever excited with sinusoidal force of the magnitude that produces small-amplitude vibrations and which frequency f_e is 10 times (2.13 (top)) and 100 times (Fig. 2.13 (middle)) larger than fundamental frequency of the microcantilever f_1 in the presence of squeeze-film damping modeled with NRE. These figures clearly indicate that the pressure lags behind the sinusoidal velocity and that the magnitude of the phase lag increases with vibration frequency. This in turn indicates that increase in vibration frequency changes the character of squeeze-film damping phenomenon by lowering viscous damping forces and raising elastic forces. In contrast, such tendency is not observed for the analogous simulation (excitation frequency is $f_e = 100f_1$) but using model with LIRE (Fig. 2.13 (bottom)). This result is expected since LIRE neglects air compressibility effects. Here pressure and velocity are approximately in phase, even we may notice that pressure somewhat leads the velocity.

2.3 Finite Element Modeling and Simulation of Electrostatic–Structural Interaction

2.3.1 Model Formulation

Electrostatic–structural interaction in a microswitch (or in any other vertically-moving MEMS actuator lying opposite to an actuation electrode) manifests as deflection of the microcantilever under the influence of bending forces generated by applied electrostatic field. In contrast to many other MEMS actuators or sensors (e.g. microresonators and microaccelerometers), a microswitch is a pull-in type device, which means that, when actuated electrostatically, it is deflected all the way down till the bottom contact electrode (drain) and therefore has to pass from stable to unstable region of operation through pull-in instability point.

Figure 2.14 shows schematic drawing of a model that is used for static and dynamic numerical analysis of coupled electrostatic–structural problem with the purpose of investigation different pull-in characteristics. The accompanying Table 2.5 provides parameter values that are typically used in this chapter unless

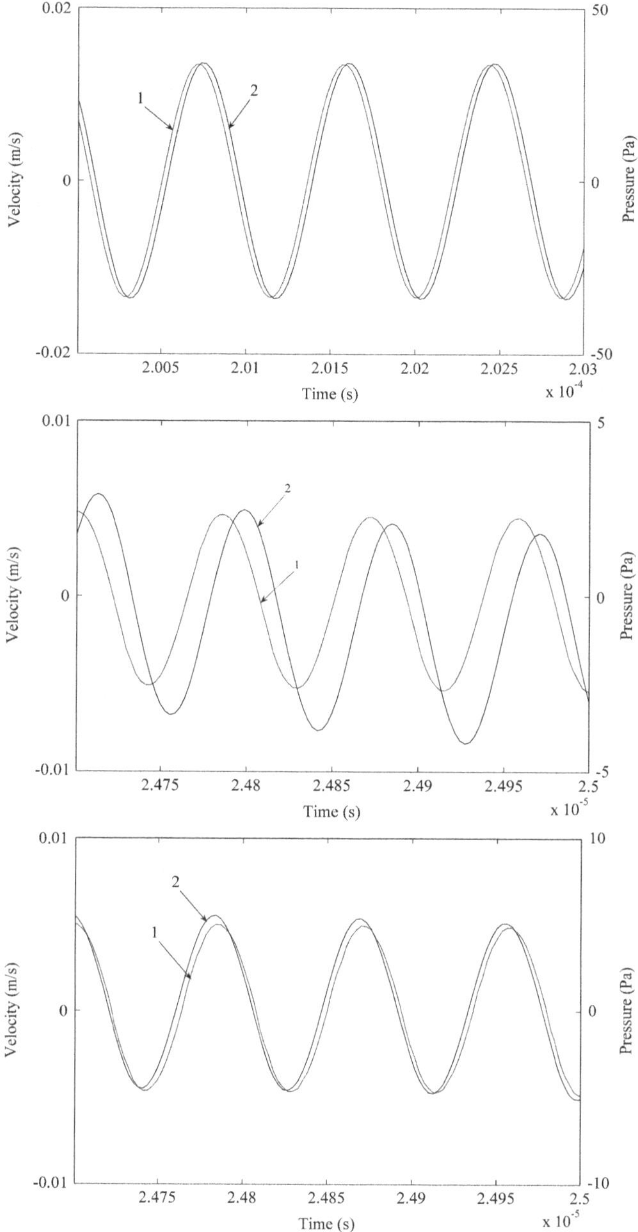

Fig. 2.13 *Top, middle*: comparison of responses of velocity (1) and pressure (2) at arbitrary midpoint at the free end of microcantilever excited with sinusoidal force of the magnitude that produces small-amplitude vibrations and which excitation frequency f_e equals to $10f_1$ (*top*) and $100f_1$ (*middle*) in the presence of squeeze-film damping modeled with NRE. *Bottom*: comparison of responses of velocity (1) and pressure (2) at arbitrary midpoint at the free end of microcantilever excited with sinusoidal force of the magnitude that produces small-amplitude vibrations and with excitation frequency $f_e = 100f_1$ in the presence of squeeze-film damping modeled with LIRE (h_0=2 μm, $p_0 = 10^5$ Pa)

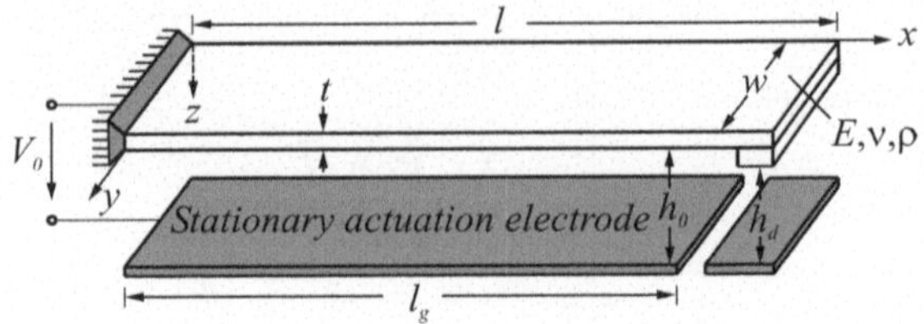

Fig. 2.14 Schematic representation of a modeled electrostatic–structural system consisting of a microcantilever actuated with forces generated by electrostatic field due to applied voltage V_0

Table 2.5 Typical parameter values used for numerical analysis of electrostatic–structural interaction

Description and symbol	Value	Unit
Length, l	117	μm
Width, w	30	μm
Thickness, t	2.0	μm
Young's modulus, E	207	GPa
Density, ρ	8902	Kg/m^3
Poisson's ratio, ν	0.31	–
Gap in gate region, h_0	Variable	μm
Gap in drain region (Tip gap), h_d	Variable	μm
Actuation voltage, V_0	Variable	V
Gate length, l_g	Variable	μm

specified otherwise. The model consists of a 3-D microcantilever which is fixed at $x=0$ and is suspended by distance h_0 over a stationary actuation electrode (gate) of length l_g. Tip gap h_d is the maximum distance that can be traversed by microcantilever tip until the touchdown occurs. Therefore this parameter is used to specify the height of imaginable contact tip ($h_0 - h_d$). Electrostatic field is generated by applying actuation voltage V_0 between the gate and the microcantilever, which electrically coincides with the source.

This model is implemented in COMSOL and it establishes two-way coupling between the microstructure deformations and the electric field. Three application modes are used: "Electrostatic", "Moving Mesh" and "Solid, Stress-Strain" for the structural part. This model involves moving boundaries therefore the program keeps track of the movements by using arbitrary Lagrangian–Eulerian (ALE) method (implemented within "Moving Mesh" application mode) and a transformation matrix, which is computed from the material displacements at the boundaries of the microstructure.

In the COMSOL model the microcantilever is placed in an air-filled chamber (surrounding box) that is electrically insulated (Fig. 2.15 (top)). The microcantilever is surrounded by a layer of air of 20 μm thickness both above and to the sides. The model uses symmetry on the z–x plane at $y = 0$ and in order to compute the electrostatic force, it calculates the electric field in this surrounding air. For the insulated surfaces of the chamber we apply boundary condition $\mathbf{n} \cdot \mathbf{D} = 0$, which means that the normal component of the electric displacement ($\mathbf{n} \cdot \mathbf{D}$) is zero (this corresponds to zero surface charge at the boundary). Bottom side of the chamber

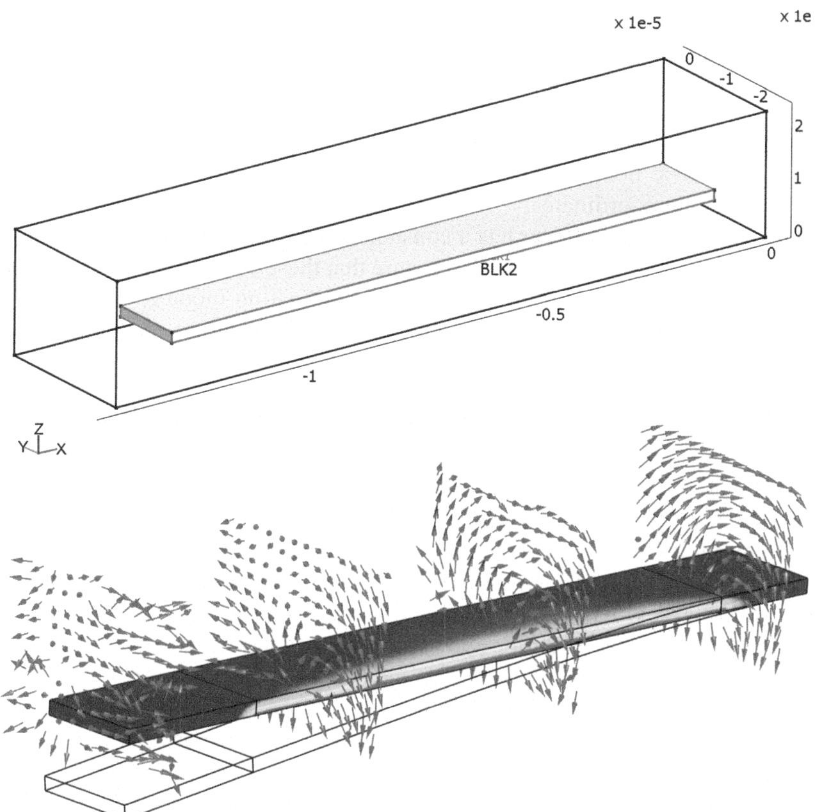

Fig. 2.15 *Top*: COMSOL finite element model of microcantilever under electrostatic actuation. *Bottom*: visualization of steady-state solution for the microcantilever of $l = 117$ μm ($l_g = 80$ μm) due to applied actuation voltage $V_0 = 67$ V. The deformed microcantilever is delineated by black edges and is hollow. Non-deformed microcantilever geometry is used to indicate distribution of electric potential. Applied electrostatic boundary conditions in the model correspond to the actual configuration of fabricated microswitches. Therefore electric potential is applied only in the limited region of microcantilever (corresponding to gate length l_g and its relative position). This region is marked by edges and has brighter color distribution (representing electric potential) in-between. Array of dark arrows show the electric field

has an electrode (corresponding to gate) that is grounded therefore its boundary condition is $V = 0$. The bottom side of the microcantilever is assumed to be coated with a thin conductive layer (corresponding to source), which is placed at a positive potential therefore its boundary condition is $V = V_0$, which specifies the actuation voltage V_0 connected to this electrode. An electrostatic force generated by an applied potential difference between the source and the gate electrodes makes the microcantilever to bend towards the grounded electrode below it. As the micro-cantilever bends, the geometry of the air changes, thus continuously changing the electric field between the electrodes. ALE method takes this air geometry changes into account when computing the potential field. The model solves the following electrostatic equation in the air domain:

$$-\nabla \cdot (\varepsilon \nabla V) = 0, \tag{2.13}$$

where V – electric potential, ε – permittivity. Laplace operator ∇ is taken with respect to moving coordinates.

Because the microcantilever has a constant potential, the electrostatic forces are always perpendicular to its surface. To ensure that these forces have this orientation, the model uses the ability of "Electrostatics" application mode to calculate Maxwell's surface stress tensors on boundaries in order to define the electrostatic force:

$$F_{es} = -\frac{1}{2}(\mathbf{E} \cdot \mathbf{D})\,\mathbf{n} + (\mathbf{n} \cdot \mathbf{E})\,\mathbf{D}^T, \tag{2.14}$$

where **E** and **D** are the electric field and electric displacement vectors, respectively, and **n** is the outward normal vector of the boundary. This force is always oriented along the normal of the boundary.

The underlying method that is used to compute microcantilever deformations due to applied voltage is based on so-called "sequential" (also called "load vector") coupling between electrostatic and structural domains. In this method the applied voltage is gradually increased and for each voltage value the coupled electrostatic–structural problem is solved iteratively, i.e. this method switches between separate single domain models in each load step in order to arrive at converged solution. The first model solves for the electrostatics around the undeformed microcantilever and then calculates the resultant distributed force acting on the microcantilever. A second model is initially associated with the undeformed microcantilever. COMSOL calculates the microcantilever deflection caused by electrostatic forces. The scripting language is used to create a new space around the deformed microcantilever, and electrostatic fields and forces are recalculated. These recalculated forces are again applied to the microcantilever to calculate a new deflection and so on. In this way the scripting language iterates between electrostatic and structural domains to arrive at a converged solution, providing results about deflections, stresses, electrostatic field and capacitance.

2.3.2 Numerical Analysis of Microswitch Operational Characteristics

Developed COMSOL model may be used for numerical analysis of different performance characteristics of electrostatically actuated MEMS devices including microswitches. The most important characteristics of microswitches are pull-in voltage V_{PI}, pull-in distance h_{PI}, pull-in time t_{PI} and pull-out time t_{PO}.

2.3.2.1 Static Simulations

Pull-in voltage V_{PI} is defined as the actuation voltage at which the resulting electrostatic force overcomes the elastic restoring force of the microcantilever and it abruptly collapses onto the bottom electrode. Its accurate determination is essential in the design process in order to determine operational performance of electrostatic microdevices. It can also be used to determine the material properties, such as the Young's modulus or the residual stresses associated with microfabricated thin films [15, 16]. Pull-in voltage is one of the most important microswitch operational parameters since it indicates the smallest voltage that can close the microswitch. The associated parameter is pull-in distance – maximum range of travel achieved by microcantilever before the start of unstable part of deflection due to the pull-in instability phenomenon.

Model verification. A series of static simulations were carried out using the developed COMSOL model in order to determine values of pull-in voltage for microswitches with dimensions and configuration of actuation electrode (their size, location) corresponding to those of actual fabricated devices: l =67, 87, 117 μm and l_g = 30, 50, 80 μm respectively, h_0 = 2 μm, h_d = 1 μm, while other analysis parameters are the same as listed in Table 2.5. Parametric solver is used for this type of numerical analysis – the program scans over a range of actuation voltages V_0 in predefined steps and thereby computes the microcantilever deflection corresponding to each voltage value. When the program tries to solve for a voltage level higher than the pull-in voltage, the solution ceases to converge since ALE technique in static mode cannot handle unstable part of microcantilever deflection corresponding to microcantilever collapsing to the bottom electrode. Thus the static simulation stops at the last converged solution, which in turn is used to determine the pull-in parameters. Figure 2.15 (bottom) provides a visualization of this steady-state simulation for one particular case. Illustration of numerical results obtained from this simulation is presented in Fig. 2.16 and allows to estimate both the pull-in voltage V_{PI} and the pull-in distance h_{PI}. As the figure reveals, the lower end of the curves is at $y \approx 0.7$ μm, which in turn indicates the value of pull-in distance and is approximately equal to $h_{PI} \approx 1/3h_0$, which corresponds to the value provided by a classical parallel-plate actuator model. Predicted pull-in voltages are compared against results of electrical measurements (Table 2.6).

This comparison indicates that simulation results are on average by 10.1% larger than experimental ones. This discrepancy is not surprising because there are a number of MEMS-specific error sources, which may reduce the accuracy of results from the COMSOL model:

(a) Uncertainty in material properties. The model uses Young's modulus of bulk nickel (207 GPa [1]), however, it is well-known that material properties of microfabricated materials can strongly deviate from bulk-material values and are very dependent on particular fabrication process as well as its technological parameters [3, 17]. For example, mechanical properties of electroplated nickel strongly depend on current density used for deposition [18]. In literature we may

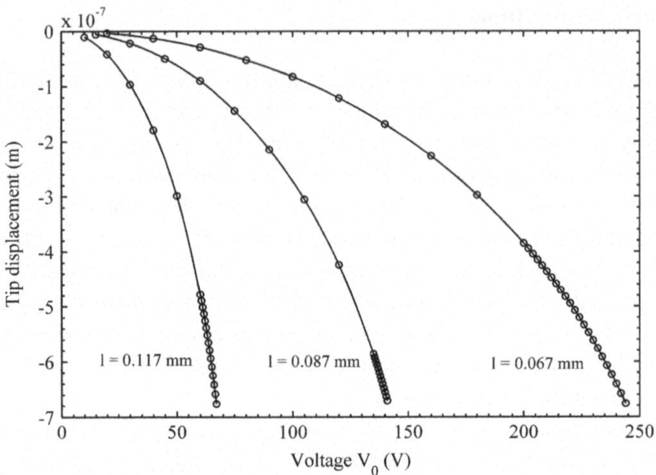

Fig. 2.16 Vertical microcantilever tip displacement at different actuation voltages V_0 for microcantilevers of different length ($h_0 = 2$ μm)

Table 2.6 Comparison of predicted pull-in voltages of fabricated microswitches with the values measured by using electrical setup with microscope-based probe station

Length of microcantilever (μm)	Simulated pull-in voltage V_{PI} (V)	Measured pull-in voltage V_{PI} (V)	Relative error (%)
67	244.70	226.70	7.94
87	141.10	128.40	9.89
117	67.30	59.90	12.35

find large variations in experimental values of Young's modulus of thin-film nickel from different authors: e.g. in [19] – 202 GPa, [3] – 200 GPa, [20] – 176 GPa, [21] – 160 GPa.

(b) Uncertainty in geometric dimensions. V_{PI} is especially sensitive to variation of microcantilever thickness t since its stiffness is proportional to t^3, and the thickness is that particular parameter of the microcantilever which is highly prone to variations because of microfabrication errors associated with changes in nickel electroplating duration and current density.

(c) Structural and anchoring idealizations. In the COMSOL model the actual microcantilever of the fabricated microswitches is approximated by a homogeneous microstructure having ideal cantilever-type fixing (anchoring) conditions. However it is well known in MEMS community that fixing conditions of microfabricated beams at the anchors usually deviate from ideal boundary conditions. Real, non-ideal anchors have either so-called step-up or cup-shaped profile that allow for more rotation and stress absorption, which in turn results in increased anchor compliance and therefore reduces microcantilever stiffness [22, 23]. Consequently this leads to lower experimental values of pull-in voltages.

With the purpose of *model verification*, multiple static simulations were carried out with COMSOL model. Predicted V_{PI} were compared with the values calculated by using analytical parallel-plate actuator model and a few of the most popular and well-known improved and semi-empirical 1-D models for electrostatic actuators:

Analytical parallel-plate actuator model [2]:

$$V_{PI} = \sqrt{\frac{8k_m}{27\varepsilon_0 l_g w} h_0^3}, \quad k_m = \tfrac{2Ewt^3}{3l^3}. \tag{2.15}$$

Improved parallel-plate actuator model [24]:

$$V_{PI}^* = 0.5477\sqrt{\frac{Et^3 h_0^3}{\varepsilon_0 l^4}}. \tag{2.16}$$

S.D. Senturia's semi-empirical model [15]:

$$V_{PI}^{**} = \sqrt{\frac{4c_1 B}{\varepsilon_0 l^4 c_2^2}}, \; c_1 = 0.07, \; c_2 = 1.00, \; B = \hat{E}th_0, \; \hat{E} = \frac{E}{1-\nu^2}, \tag{2.17}$$

where k_m is the microcantilever stiffness, $\hat{E}$ – effective modulus of elasticity for a plate (analyzed microcantilever is considered to be a plate since its width w is more than 5 times larger than its thickness t, which is a criterion indicated by S. Timoshenko [15, 23]).

This analysis was performed for typical microcantilevers with $l = l_g = 67$, 87, 117 μm, $h_0 = h_d = 2$ μm. Table 2.7 summarizes the results, which indicate that the simplest parallel-plate actuator model significantly underestimates the value of pull-in voltage, while values of V_{PI} from other two models are very close to the predictions of the COMSOL model. Consequently, this comparative analysis demonstrates that the developed model is capable of delivering accurate results.

Furthermore, for model validation purposes it is also worthwhile to compare simulated values of pull-in voltages with reliable experimental results of other authors for similar microstructures. Measurement data provided by R.K. Gupta from MIT [16] is particularly suited for this task since it was obtained from experiments with

Table 2.7 Comparison of values of pull-in voltages obtained with COMSOL and other models

Length of microcantilever (μm)	Theoretical pull-in voltage V_{PI} (V)			
	Authors' COMSOL model (3-D)	Analytical parallel-plate actuator model (1-D) [2]	Improved parallel-plate actuator model (1-D) [24]	S.D. Senturia's semi-empirical model (1-D) [15]
67	154.00	121.11	149.24	151.66
87	87.00	71.83	88.51	89.95
117	48.20	39.71	48.94	49.73

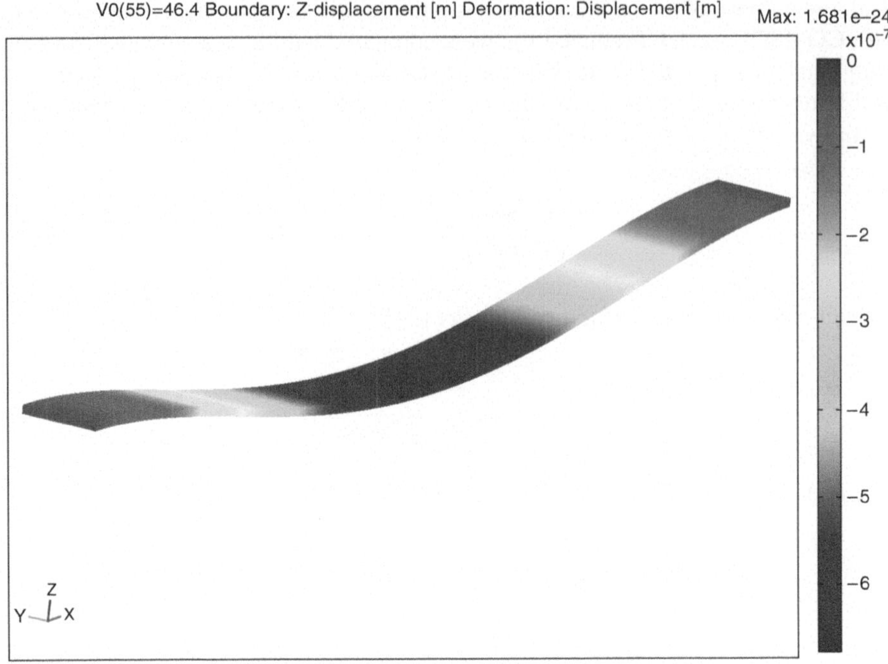

Fig. 2.17 Visualization of deflected fixed–fixed microbeam (l = 300 μm) due to applied voltage $V_0 = 47.60$ V. Microbeam parameters used for simulation: $l = l_g = 300 \div 500$ μm, $w = 40$ μm, $t = 2.1$ μm, $h_0 = 2$ μm, $E_{Poly\text{-}Si} = 155$ GPa, $\nu = 0.23$, $\rho = 2{,}330$ kg/m^3

Table 2.8 Comparison of predicted pull-in voltages from COMSOL model for fixed–fixed microbeam with measured values determined by R.K. Gupta [16]

Length of fixed–fixed microbeam (μm)	Simulated pull-in voltage V_{PI} (V)	Experimental pull-in voltage V_{PI} (V) [16]	Relative error (%)
300	47.60	49.00	2.86
350	34.50	35.00	1.43
400	26.30	27.00	2.59
450	19.80	19.00	4.21
500	16.70	16.00	4.38

well-definedmicrostructures fabricated by means of fully-developed and accurate microfabrication technology called Multi-User MEMS Processes (MUMPs®), which together with SUMMiT® process of Sandia National Laboratories may be considered as the most renowned and ultimate platforms for MEMS fabrication in the world [3]. Moreover, the microstructures tested by R.K. Gupta are not cantilever but fixed–fixed type therefore it is useful to check adaptability of COMSOL model for pull-in simulations on structures with different fixing conditions. Static simulations were performed on fixed–fixed microstructures with parameters taken from [16] and provided in Fig. 2.17. Obtained results (Table 2.8) indicate that pull-in voltages predicted by COMSOL model closely match R.K. Gupta's experimental data with average relative error being equal to 3.1%. This in turn confirms that the developed FE model is adequate and is suitable for numerical analysis of static pull-in parameters that are characteristic to electrostatic actuators.

Parametric analysis. Value of pull-in voltage depends on and may be controlled by a number of system parameters including: microcantilever stiffness k_m, nominal gap thickness h_0, tip gap h_d, overlap area between the gate and the microcantilever A_g and gate length l_g. A series of finite element simulations were carried out with the COMSOL model in order to conduct a parametric study for determination of dependences of pull-in voltage on aforementioned parameters. Figure 2.18 presents the evolution of pull-in voltage with respect to a wide range of values of microcantilever stiffness k_m. The purpose of this numerical analysis was to modify only structural characteristics of the microsystem therefore value of k_m was changed by varying microcantilever thickness t because variation of length or width also changes magnitude of electrostatic forces since these geometric parameters simultaneously affect not only structural stiffness but also the size of overlap area A_g and the relative position of gate electrode. It is obvious from the obtained characteristic in Fig. 2.18 that for stiffer structure a larger voltage is needed in order to bend it. Thus, desired value of microswitch pull-in voltage may be achieved by tuning microcantilever stiffness, which may be altered by adjusting geometric dimensions of the microcantilever, modifying its shape and configuration as well as selecting different materials for fabrication.

The variation of pull-in voltage for different values of initial gap h_0 is illustrated in Fig. 2.19. In this case, only the distance between the microcantilever and the gate is modified. It means that structural characteristics of the microsystem are not modified, only the electrostatic forces are increased when the gap h_0 is reduced resulting in lower pull-in voltages. Another structural method for reducing pull-in

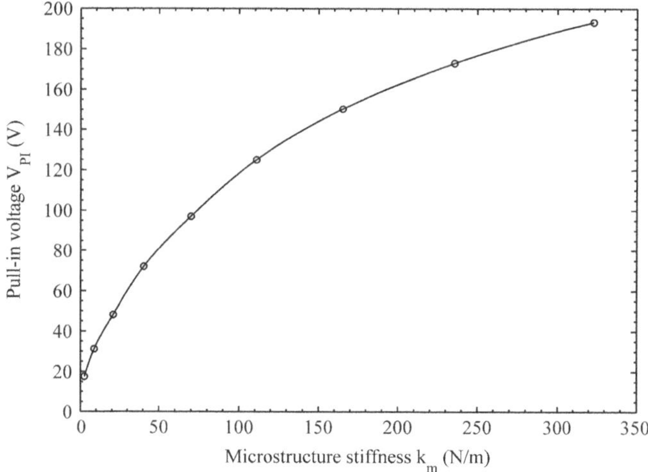

Fig. 2.18 Pull-in voltage of microswitch V_{PI} as a function of the microcantilever stiffness k_m ($l = 117$ µm, $h_0 = 2$ µm)

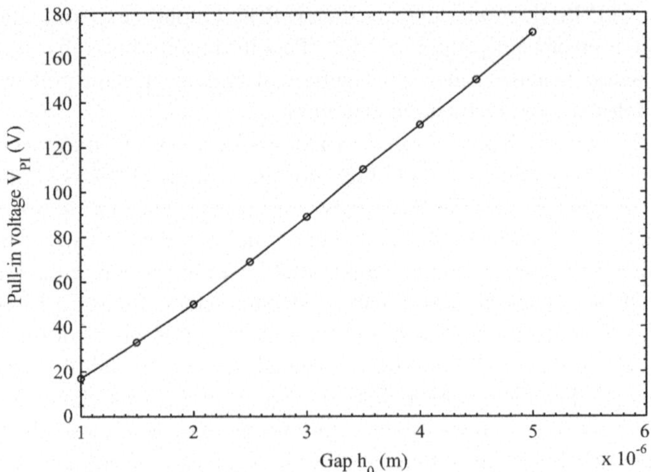

Fig. 2.19 Pull-in voltage of a microswitch V_{PI} as a function of nominal gap size h_0 ($l = 117$ μm)

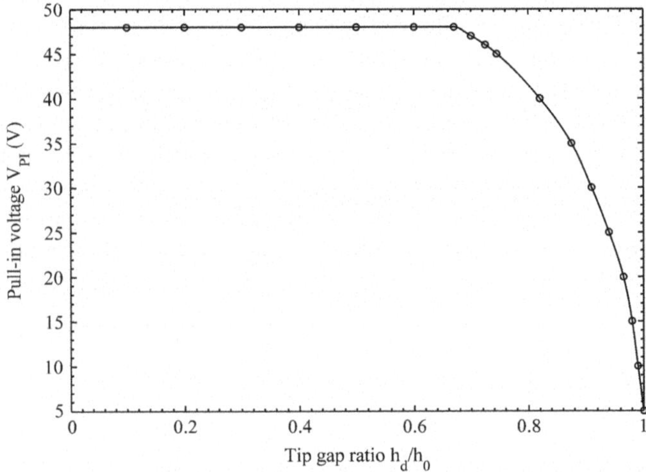

Fig. 2.20 Pull-in voltage V_{PI} as a function of tip gap ratio h_d/h_0 ($l = 117$ μm, $h_0 = 2$ μm)

voltage is to introduce miniature contact tips (bumps) on the bottom side of the free end of the microcantilever thereby reducing the distance that the microcantilever has to be deflected until it touches the bottom electrode. The influence of height of contact tips on the value of pull-in voltage is illustrated in Fig. 2.20 in the form of relationship between V_{PI} and tip gap ratio h_d/h_0. The shape of this plot leads to a conclusion that due to the pull-in instability phenomenon the influence of contact tips starts to manifest only when their height exceeds $2/3h_0$. It is obvious that the tips which height is larger than $2/3h_0$ have a pronounced effect in reducing pull-in voltage. The magnitude of generated electrostatic forces may be regulated

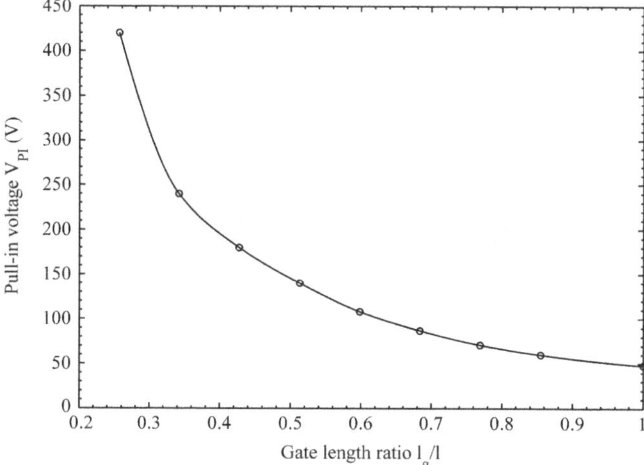

Fig. 2.21 Pull-in voltage of a microswitch V_{PI} as a function of gate length ration l_g/l ($l = 117$ μm, $h_0 = 2$ μm)

by modification of the size of the gate electrode thereby changing the overlap area $A_g = l_g w$. Thus, one of the ways to achieve this is to change gate length l_g. The effect of different values of l_g on V_{PI} is demonstrated in Fig. 2.21, which indicates that reduction of overlap area dramatically raises the pull-in voltage leading to unacceptably high values for $l_g = 1/2l$ and less.

2.3.2.2 Dynamic Simulations

Transient simulations are important in order to describe dynamic behavior of the microswitch and determine its switching speed, which is characterized by two parameters: (1) pull-in time t_{PI}, which is the time required for the microcantilever to be deflected all the way down till the bottom electrode when the pull-in or larger actuation voltage is applied and (2) pull-out time t_{PO} – time required for the microcantilever to return from its deflected down-state position with zero air-gap to initial undeformed up-state position. A series of transient simulations were performed in order to estimate these parameters and study their dependence on various system parameters. Damping was neglected in this analysis.

The aim of the initial finite element simulations was to determine transient microcantilever tip displacements for different actuation voltages V_0. It is obvious from the results presented in Fig. 2.22 that pull-in time strongly depends on the applied voltage since the larger the voltage, the stronger the electrostatic force. It is more difficult to validate the applicability of the COMSOL model for making t_{PI} predictions since there were no possibilities to perform switching speed measurements on fabricated microswitches. Furthermore, unfortunately there are no specially-derived accurate

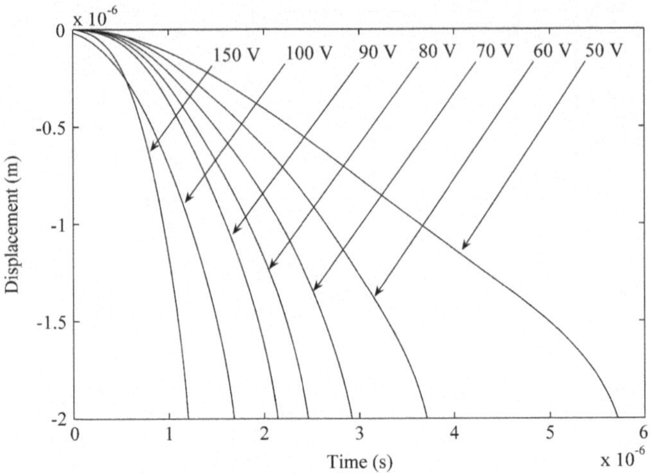

Fig. 2.22 Microcantilever tip displacement variation in time after the microstructure has been subjected to different actuation voltages V_0 ($l = 117$ μm, $h_0 = 2$ μm)

Table 2.9 Comparison of predicted pull-in times for different values of applied voltage with values calculated by using approximate expression that was derived on the basis of lumped-parameter parallel-plate actuator model

Applied voltage V_0 (V)	Theoretical pull-in time t_{PI} (μs)	
	Authors' COMSOL model (3-D)	Analytical parallel-plate actuator model (1-D) [2]
50	5.71	5.00
60	3.70	4.15
70	2.92	3.56
80	2.46	3.12
90	2.05	2.77
100	1.68	2.49
150	1.20	1.66

expressions for pull-in time (contrary to the case of presented improved models for V_{PI}). Only the following approximate expression for t_{PI}, derived on the basis of analytical parallel-plate actuator model, is available in the literature [2]:

$$t_{PI}^{*} \approx 3.67 \frac{V_{PI}}{V_0 2\pi f_0}, \tag{2.18}$$

where f_0 – first natural (fundamental) frequency of the microcantilever.

Comparison of simulation and analytical results for pull-in times, presented in Table 2.9, indicates that they agree qualitatively. Quantitatively, simulated values differ from analytical ones by an average of 21.5%, however it should be noted that they match each other better than the equivalent comparison of pull-in voltages provided in Table 2.7 (the average relative error in this case is equal to 23.2%).

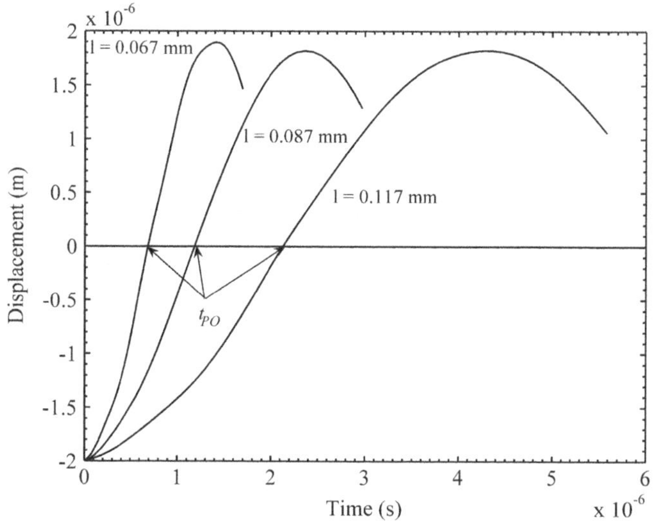

Fig. 2.23 Fragments of free vibration curves of microcantilever of different length l

In order to estimate pull-out time t_{PO} for microcantilevers of different length, the static problem is first solved in COMSOL in order to deflect the microcantilever tip until the bottom electrode thus achieving zero air-gap condition that corresponds to microswitch being in closed down-state position. Then this static solution is used as initial condition for transient analysis. In this stage of analysis the microcantilever is released and oscillates freely around its equilibrium position. The fragments of the resulting free vibration curves are presented in Fig. 2.23. The first zero crossing of the curves with x-axis is used as a measure of pull-out time as it is indicated in the figure. It should be noted that the latter approach for determination of t_{PO} provides only a rough estimate of the lowest possible value of this parameter under the most favorable damping conditions. In reality the microcantilever after release will oscillate freely a number of periods depending on the actual damping of the system. The pull-out times estimated from the simulations are 2.18 μs for microcantilever of $l = 117$ μm, 1.22 μs – for $l = 87$ μm and 0.7 μs – for $l = 67$ μm. Naturally, the pull-out time varies proportionally to the fundamental frequency of the microcantilever f_0, and therefore may be reduced by increasing structural stiffness. Comparison of computed pull-in and pull-out times indicates that the latter are shorter than the former except for the case when microswitch is actuated by a very high voltage V_0. It should be pointed out that it is the pull-in time that is commonly mentioned in the literature as a limiting component of the switching speed of microswitches and therefore deserves particular attention. For this reason a number of transient simulations were carried out in order to identify and study possible means of controlling this characteristic. It is possible to reduce pull-in time t_{PI} by applying actuation voltage V_0 that is higher than the value of pull-in voltage V_{PI} for the given microswitch. After performing dynamic simulations their results were summarized in the plot of $t_{PI} = f(V_0/V_{PI})$, which is illustrated in Fig. 2.24. It indicates that

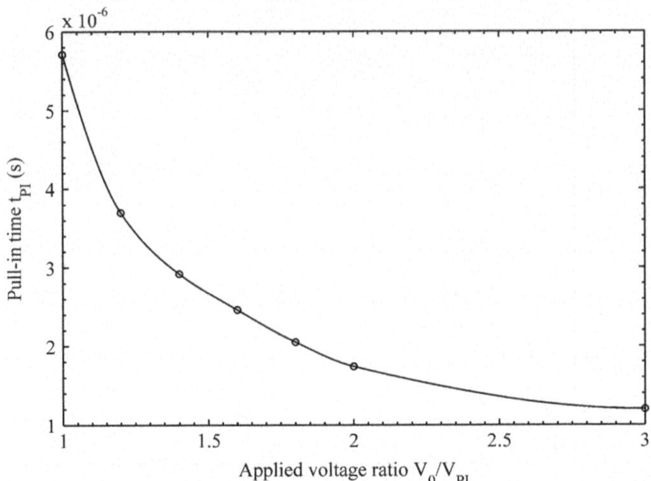

Fig. 2.24 Pull-in time t_{PI} as a function of applied voltage ratio V_0/V_{PI} ($V_{PI} = 48.2$ V, $l = 117$ μm)

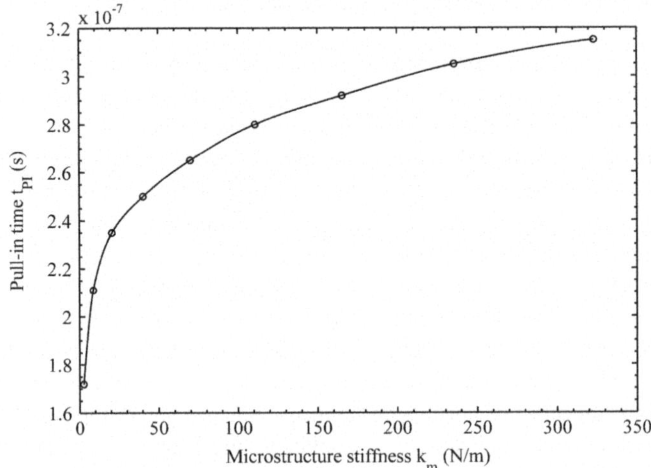

Fig. 2.25 Pull-in time of a microswitch t_{PI} as a function of microcantilever stiffness k_m ($l = 117$ μm, $h_0 = 1$ μm, $V_0 = 250$ V)

applied voltage has pronounced effect on the value of pull-in time: increasing V_0 by 3 times allows to diminish t_{PI} nearly 5 times.

There are several structural means available that may be applied for control of t_{PI}. One of them is modification of microcantilever stiffness k_m. Its influence on value of t_{PI} is demonstrated in Fig. 2.25. During simulations the stiffness was varied by changing thickness of the microcantilever. It is obvious from the presented results that the sensitivity of pull-in time to change of k_m is not uniform throughout considered stiffness range of $3 \div 350$ N/m. The plot indicates that t_{PI} increases with stiffness more rapidly in the beginning – in the region of approximately $3 \div 70$ N/m. The range

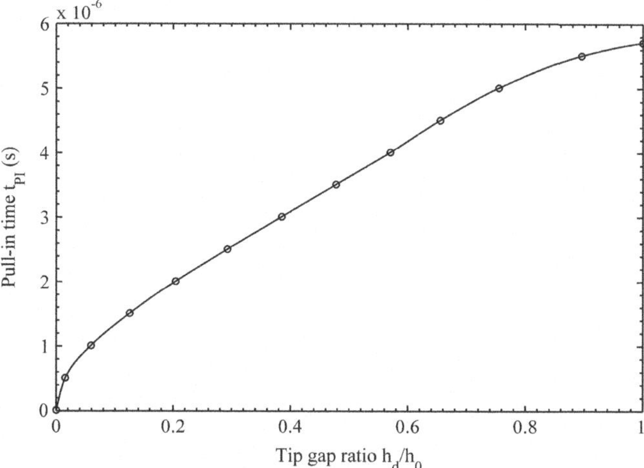

Fig. 2.26 Pull-in time of t_{PI} as a function of tip gap ratio h_d/h_0 ($l = 117$ μm, $h_0 = 2$ μm, $V_0 = 50$ V)

of 3 ÷ 70 N/m corresponds to thickness range of 1 ÷ 3 μm for the analyzed microcantilever. Moreover, this range includes values of structural thickness that are commonly encountered in many different MEMS devices. Consequently, this range is the most suitable for the adjustment pull-in time during device design stage. On the other hand, in this range pull-in voltage V_{PI} is also the most sensitive to changes of k_m. Thus, it is a challenging task to achieve low pull-in voltage and high switching speed simultaneously since, in general, increase in structural stiffness leads to shorter pull-in and pull-out times but to larger pull-in voltage. Therefore a compromise is usually required between these two characteristics during the microswitch design in order to obtain the best device performance.

Another structural approach for reducing pull-in time of a microswitch is to use contact tips at the end of the microcantilever. Figure 2.26 presents the dependence of $t_{PI} = f(h_d/h_0)$, which allows to judge about influence of tip height on value of pull-in time.

Results of numerical analysis of squeeze-film damping presented in Section 2.2 demonstrated that for typical dimensions of microdevices the damping can be very high and reduce the quality factor Q of the system below one. The results of this analysis may be applied for investigation of influence of air damping on dynamic pull-in characteristics. Quality factors extracted from frequency response analysis of microcantilever under the effect of squeeze-film damping were used in order to add viscous damping to the COMSOL model by means of Rayleigh damping approach:

$$[C] = \alpha_{dM}[M] + \beta_{dK}[K], \tag{2.19}$$

where [M], [C], [K] are mass, damping and stiffness matrices respectively in the equation of motion of the dynamic system, α_{dM}, β_{dK} are mass and stiffness damping parameters respectively.

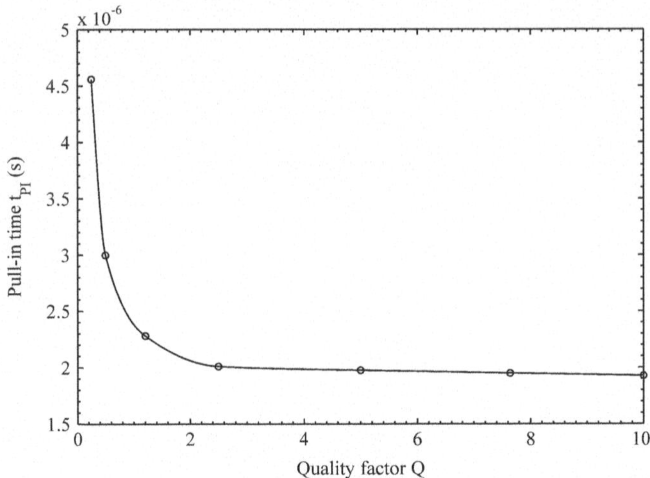

Fig. 2.27 Pull-in time t_{PI} as a function of quality factor Q ($V_0 = 100$ V, $l = 117$ μm, $h_0 = 2$ μm)

The quality factors were recalculated to the β_{dK} by means of the following expression and inserted into the COMSOL model:

$$\beta_{dK} = \frac{1}{Q\omega_0}. \tag{2.20}$$

The model with damping was used to determine pull-in times for different values of Q. Results presented in Fig. 2.27 reveal that below $Q = 0.5$ pull-in time increases dramatically, while for $Q > 3$ pull-in time does not change any more. Thus, with respect to switching speed, it is advantageous to have a structure with $0.5 \leq Q \leq 2$ since: (a) $Q > 2$ would result in a high value of pull-out time due to long settling of the microcantilever when the microswitch is released from its down-state position, (b) reduction of pull-in time for $Q > 2$ is very small.

2.4 Finite Element Modeling and Simulation of Vibro-Impact Interaction

2.4.1 Model Formulation

Vibro-impact interaction in a microswitch manifests as a contact bouncing, i.e. parasitic process of repetitive bouncing of the microcantilever tip after it has impacted bottom electrode (drain) during microswitch closure. Figure 2.28 shows a schematic drawing of a model that is used for numerical analysis of contact bouncing effects. The model consists of a 3-D microcantilever, which is fixed at $x = 0$ and is suspended by distance h_0 over a stationary structure, which also

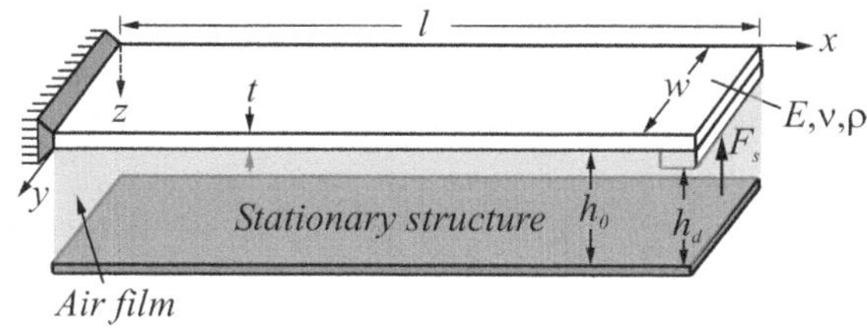

Fig. 2.28 Schematic representation of a modeled fluidic–structural system consisting of a microcantilever under the effect of squeeze-film damping and adhesive–repulsive force F_s used for modeling of vibro-impact interaction between the microcantilever and stationary structure

Table 2.10 Typical parameter values used for numerical analysis of vibro-impact interaction

Description and symbol	Value	Unit
Length, l	117	μm
Width, w	30	μm
Thickness, t	2.0	μm
Young's modulus, E	207	GPa
Density, ρ	8902	Kg/m^3
Poisson's ratio, ν	0.31	–
Gap, h_0	variable	μm
Tip gap, h_d	0	μm
Shear force, F_s	–	N

represents imaginary contact plane. Parameters listed in Table 2.10 are typically used in this chapter unless specified otherwise.

It is assumed that both structures are surrounded by the air therefore the model may be used to solve contact problem taking into account also the influence of squeeze-film damping. Tip gap h_d is the distance between microcantilever tip and contact plane. Tip/drain vibro-impact interaction is modeled by a special adhesive-repulsive contact model, whose repulsive contribution is described by a classical linear elastic force and adhesive contribution is represented by an attractive van der Waals force component. The expression for attractive van der Waals component is derived from the following relationship for van der Waals force [25]:

$$F_c = \frac{A_H A_c}{6\pi D^3} = \frac{C_{vdW}}{D^3}, \tag{2.21}$$

where A_c is contact area, D – the sum of the RMS roughness of the interacting surfaces, A_H – Hamaker constant, which characterizes the van der Waals attraction of solids across a liquid or gas and depends on the materials and intervening medium (typically it is assumed to be about 1×10^{-19} J, but for interactions between metals its theoretical values calculated from Lifshitz theory is about 4×10^{-19} J [25]).

This adhesive-repulsive contact model is implemented in COMSOL package by using shear force $F_s(t)$ that is applied on the lower edge at free end of the microcantilever as indicated in Fig. 2.28:

$$F_s(t) = \begin{cases} \frac{C_{vdW}}{[h_d - z_l(t)]^3} & for\ z_l(t) < h_d - \xi_0 \\ \frac{C_{vdW}}{\xi_0^3} - k_p[z_l(t) - (h_d - \xi_0)] & for\ z_l(t) \geqslant h_d - \xi_0 \end{cases}, \tag{2.22}$$

where ξ_0 is the interatomic distance at equilibrium in the interface between microcantilever tip and stationary structure (its value is based on lattice spacing and is assumed to be 1 nm), $z_l(t)$ – displacement of microcantilever tip, k_p – "penalty" stiffness coefficient.

In this contact model tip/drain interaction is modeled: (a) before contact (when $z_l(t) < h_d - \xi_0$) – by attractive van der Waals force component, that is proportional to constant parameter $C_{vdW} = A_h A_c / 6\pi$, and decays with the third power of the separation distance $h_d - z_l(t)$, (b) upon contact (when $z_l(t) \geq h_d - \xi_0$) – by additional repulsive linear elastic force, which is proportional to the stiffness k_p and grows linearly with the penetration distance.

The proposed contact model is adaptive because it allows to formulate different contact conditions, starting from purely repulsive contact condition without contribution of adhesion (by setting $C_{vdW} = 0$) to a mixed adhesive-repulsive contact condition. Thus, by changing value of parameter C_{vdW} it is possible to obtain and analyze different contact conditions ranging from fully-repulsive to adhesive-repulsive ones.

2.4.2 *Numerical Analysis of Contact Bouncing*

In order to study contact bouncing effects, microcantilever tip is forced to impact stationary structure in the following way:

(a) Lower edge at the free end of the microcantilever is displaced upwards by distance z_0. This is accomplished by solving static problem in COMSOL;
(b) The result of the static analysis is used as initial condition for subsequent transient analysis. In this analysis the microcantilever is released from its deformed position at the moment $t = 0$ and is further forced to impact the imaginary contact plane.

During simulations the distance h_d was equal to zero. It means that in its initial condition, before static analysis, the microcantilever "rests" on the contact plane. For this case, the expression for $F_s(t)$ in Eq. (2.22) is modified by replacing h_d with z_0 and the resulting final expression is inserted into the COMSOL model via graphical user interface of the software.

The actual value of "penalty" stiffness coefficient k_p is not known a priori. For the simulations performed in this work, its value was specified taking into account a rule

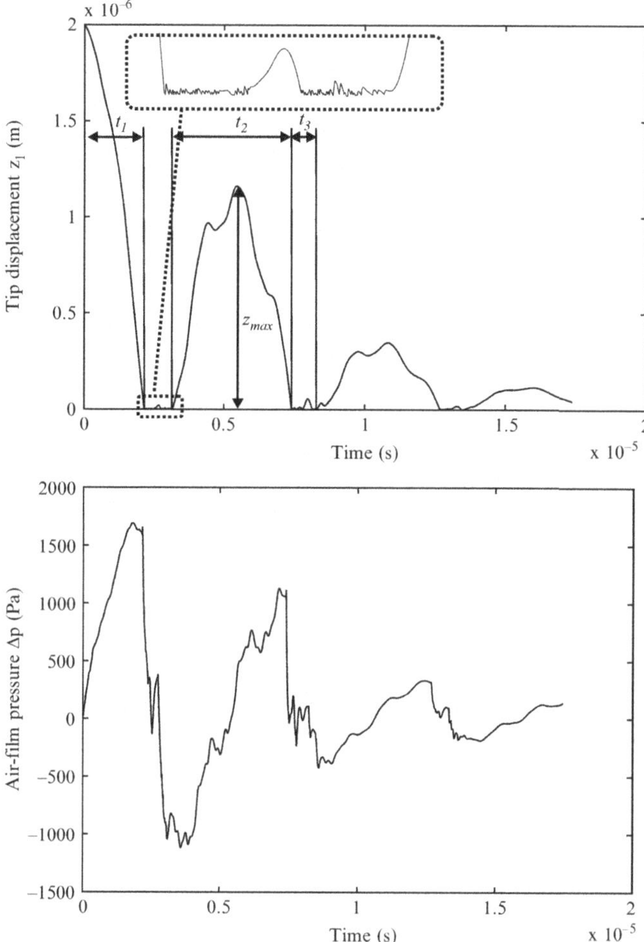

Fig. 2.29 *Top*: typical curve of free impact vibrations - tip displacement variation in time after free end of the microcantilever has been deflected upwards by $z_0 = 2$ μm and released thereby forcing it to come into contact with the stationary structure and bounce afterwards ($l = 117$ μm, $C_{vdW} = 10^{-30}$ Nm^3). *Bottom*: the corresponding variation of air-film pressure Δp at arbitrary mid-point located near the free end of microcantilever

of thumb that the value of k_p should be at least 100 times larger than the highest stiffness coefficient of other system components. The spring constant of the typical microcantilever considered in this work is $k_m = 20$ N/m. Initial vibro-impact simulations revealed that numerical convergence is best achieved when $k_p = 10^4 \div 10^7$ N/m.

Figure 2.29 (top) illustrates a typical curve of free impact vibrations, which was obtained by applying above-described procedure of numerical analysis. The enlarged area in the figure shows in detail the phenomenon of chatter that occurs after the first contact of the microcantilever tip with the stationary structure. The most important characteristics of the curve of free impact vibrations are: t_1 – time to the first contact, t_2 – bounce duration, t_3 – impact duration, $t_4=t_2+t_3$ – period of free impact vibrations, z_{max} – bounce amplitude. Meanwhile, Fig. 2.29 (bottom) presents the temporal variation of additional air-film pressure Δp due to the squeezed film effect at arbitrary point, which is located in the middle of microcantilever towards its free end. This plot indicates that when microcantilever approaches the contact plane in the beginning, the pressure is gradually increased above atmospheric value by $\Delta p \approx 1{,}700$ Pa. After the first contact occurs, the microcantilever bounces off and retreats, and in the meantime, the pressure reduces below atmospheric value by more than $\Delta p \approx 1{,}000$ Pa thus indicating the effect of decompression in the gap.

As have already been demonstrated in Section 2.2, the magnitude of squeeze-film damping in the system is determined by the ambient pressure p_0 and air-film thickness h_0. In order to study influence of air damping in the case of vibro-impact interaction of the moving microcantilever with the stationary structure, two series of transient simulations were carried out. In the first one, pressure p_0 was varied and the rest of system parameters were kept constant. Figure 2.30 illustrates numerical results, which were obtained with initial deflection of $z_0 = 2$ μm. The presented plots indicate that for the case of $h_0 = 2$ μm and $p_0 = 10 \div 100$ kPa, ambient air generates an appreciable amount of damping. The relative reduction of the first bounce amplitude z_{max} at $p_0 = 10$ kPa and $p_0 = 100$ kPa with respect to vacuum conditions is equal to 6% and 15% respectively. While, in pressure range of $p_0 = 0.01 \div 1$ kPa the obtained values of z_{max} are close to each other and indicate insignificant difference in comparison to vacuum conditions.

In the second series of transient simulations ambient pressure was fixed at $p_0 = 100$ kPa, while air-film thickness was varied in the range $h_0 = 0.75 \div 2.5$ μm.

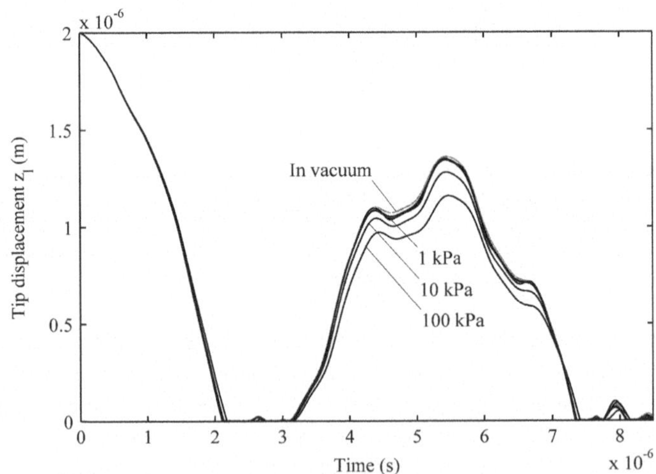

Fig. 2.30 Tip displacement variation in time $z_l(t)$ both in vacuum and in air with pressure p_0 after free end of the microcantilever has been deflected upwards by $z_0 = 2$ μm and released thereby forcing it to come into contact with the stationary structure and bounce afterwards ($l = 117$ μm, $h_0 = 2$ μm, $C_{vdW} = 10^{-30}$ Nm^3)

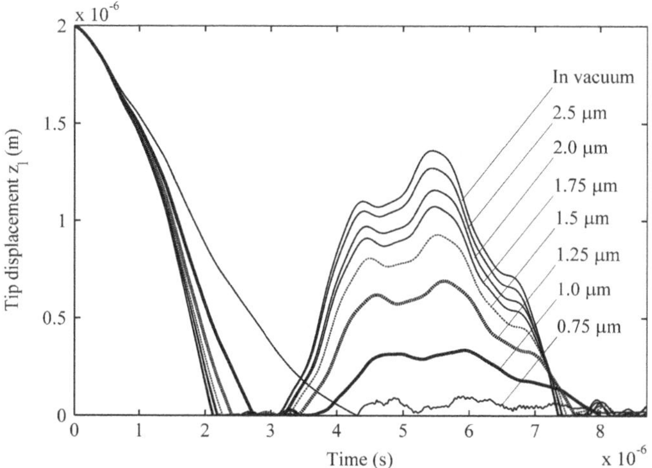

Fig. 2.31 Tip displacement variation in time $z_l(t)$ for different h_0 both in vacuum and in air with pressure $p_0 = 100$ kPa after free end of the microcantilever has been deflected upwards by $z_0 = 2$ μm and released thereby forcing it to come into contact with the stationary structure and bounce afterwards ($l = 117$ μm, $C_{vdW} = 10^{-30}$ Nm^3)

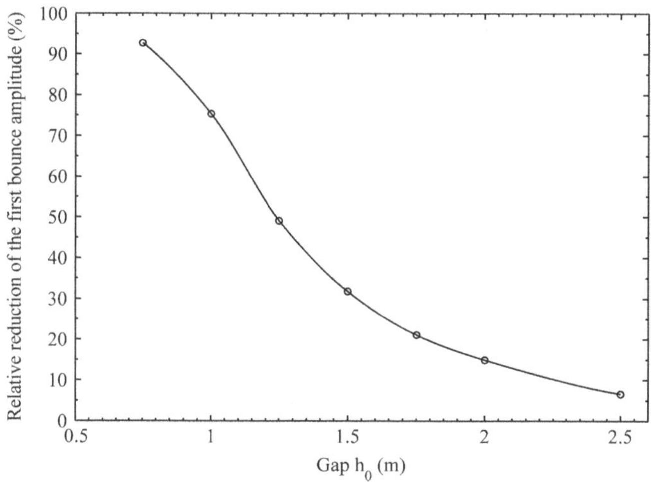

Fig. 2.32 Relative reduction of the first bounce amplitude z_{max} (with respect to value obtained under vacuum conditions) presented as a function of air-film thickness h_0 ($l = 117$ μm, $z_0 = 2$ μm, $C_{vdW} = 10^{-30}$ Nm^3)

Obtained results are summarized in Fig. 2.31 In this case the influence of squeeze-film damping on the contact bouncing behavior is much more pronounced in comparison to previous case. As Fig. 2.32 demonstrates, the amplitude of the first contact bounce reduces dramatically as the gap with air reduces in size and for the

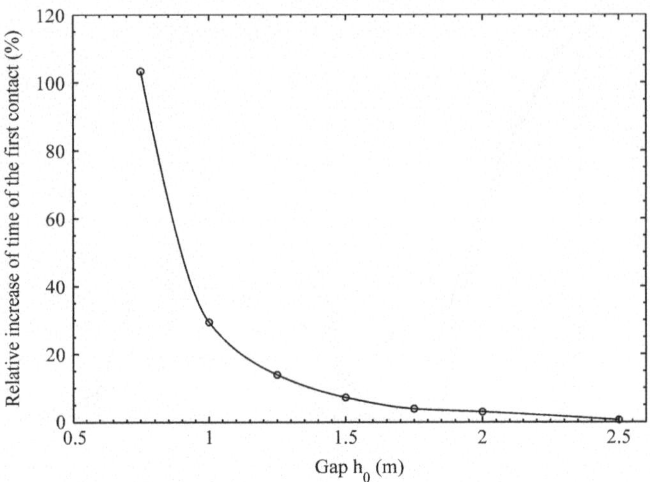

Fig. 2.33 Relative increase of time to the first contact t_l (with respect to value obtained under vacuum conditions) presented as a function of air-film thickness h_0 (l = 117 μm, z_0 = 2 μm, $C_{vdW} = 10^{-30}$ Nm³)

h_0=0.75 μm the bounce amplitude z_{max} is by more than 90% lower than the value obtained under vacuum conditions. One more obvious effect of reduced air-film thickness is the increase of the time to the first contact t_1. As the Fig. 2.33 indicates the increase of t_1 becomes particularly significant for $h_0 < 1.25$ μm.

Proposed vibro-impact model may be used to estimate contact bouncing characteristics as a function of adhesion-related parameter $C_{vdW} = A_h A_c / 6\pi$ and geometry of the microstructure. Value of contact area A_c for fabricated microswitches is calculated by taking into account total area of two contact tips, which is equal to 32 μm². However, this parameter, in general, can acquire fairly different values if we consider microdevices fabricated by different research groups. For example, in [26] contact area of proposed microswitches is as low as 0.1 μm², while for microswitches presented in [27] A_c reaches even 8,000 μm². Thus it is reasonable to use range of $A_c = 0.1 \div 10^4$ μm² for numerical analysis of contact bouncing since this range includes values of contact area that are commonly encountered in actual microswitches. If we insert these values into expression for C_{vdW} we obtain that the range of interest for adhesion-related parameter is $C_{vdW} = 10^{-27} \div 10^{-32}$ Nm³. In order to estimate the effect of C_{vdW} on dynamic response of the system, a series of simulations were performed, which results are illustrated in Fig. 2.34. We may observe that variation of this parameter does not have significant influence on the first bounce amplitude z_{max}, which is nearly coincidental for different values of C_{vdW}. However, influence of adhesion forces becomes more pronounced with each subsequent bounce. Bouncing reduces more rapidly for larger values of C_{vdW}, corresponding to larger adhesive forces.

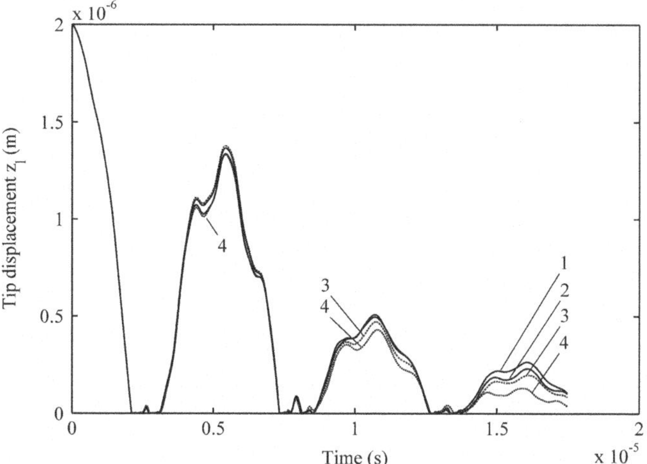

Fig. 2.34 Tip displacement variation in time $z_l(t)$ for different C_{vdW} in vacuum after the free end of microcantilever was deflected upwards by $z_0 = 2$ μm and released thereby forcing it to come into contact with the stationary structure and bounce afterwards: 1 – $C_{vdW} = 0$, 2–10^{-32} Nm^3, 3–10^{-30}, 4–10^{-27}

References

1. Nunes R, Adams JH, Ammons M et al (1990) ASM handbook, Vol. 02 – Properties and selection – nonferrous alloys and special-purpose materials. ASM International
2. Rebeiz GB (2003) RF MEMS: theory, design, and technology. Wiley-Interscience, Hoboken, NJ
3. Gad-el-Hak M (2006) MEMS handbook. CRC Press, Boca Raton, FL
4. Maluf N, Williams K (2004) An introduction to microelectromechanical systems engineering. Artech House, Boston, MA
5. Bao M, Yang H (2007) Squeeze film air damping in MEMS (review). Sensor Actuat A Phys 136:3–27
6. Lee JH, Lee ST, Yao CM, Fang W (2007) Comments on the size effects on the microcantilever quality factors in free air space. J Micromech Microeng 17:139–147
7. Hosaka H, Itao K, Kuroda S (1995) Damping characteristics of beam-shaped micro-oscillators. Sensor Actuat A Phys 49:87–95
8. Senturia SD (2001) Microsystem design. Kluwer Academic, Norwell, MA
9. Park YH, Park KC (2004) High-fidelity modeling of MEMS resonators, part I, anchor loss mechanisms through substrate. J Microelectromech S 13:238–247
10. Pelesko JA, Bernstein DH (2003) Modeling MEMS and NEMS. Chapman Hall and CRC Press, Boca Raton, FL
11. Hamrock BJ, Schmid SR, Jacobson BO (2004) Fundamentals of fluid film lubrication. Marcel Dekker, New York
12. Starr JB (1990) Squeeze-film damping in solid-state accelerometers. In: Proceedings of the IEEE solid-state sensor and actuator workshop, Hilton Head Island, SC, June 1990, pp 44–47
13. Tilmans HAC, Legtenberg R (1994) Electrostatically driven vacuum-encapsulated polysilicon resonators (part I & II). Sensor Actuat A Phys 45:67–84
14. Kelly SG (2000) Fundamentals of mechanical vibrations. McGraw-Hill, Boston, MA

15. Osterberg PM, Senturia SD (1997) M-TEST: a test chip for MEMS material property measurement using electrostatically actuated test structures. J Microelectromech S 6(2):107–118
16. Gupta RK (1997) Electrostatic pull-in test structure design for mechanical property characterization of microelectromechanical systems (MEMS). Dissertation, M.I.T.
17. Spearing SM (2000) Materials issues in microelectromechanical systems (MEMS). Acta Mater 48:179–196
18. Fritz T, Cho HS, Hemker KJ, Mokwa W, Schnakenberg U (2002) Characterization of electroplated nickel. Microsyst Technol 9:87–91
19. Mazza E, Abel S, Dual J (1996) Experimental determination of mechanical properties of Ni and Ni-Fe microbars. Microsyst Technol 2:197–202
20. Sharpe W, Lavan D, Edwards R (1997) Mechanical properties of LIGA-deposited Nickel for MEMS. In: Proceedings of international conference on solid state sensors and actuators, 1997, pp 607–610
21. Bucheit T, Christenson T, Schmale D, Lavan D (1998) Understanding and tailoring the mechanical properties of LIGA fabricated materials. In: Proceedings of MRS symposium, vol 546, 1998, pp 121–126
22. Kobrinsky MJ, Deutsch ER, Senturia SD (2000) Effect of compliance and residual stress on the shape of doubly supported surface-micromachined beams. J Microelectromech S 9:361–369
23. Lishchynska M, Cordero N, Slattery O, O'Mahony C (2005) Modelling electrostatic behaviour of microcantilevers incorporating residual stress gradient and non-ideal anchors. J Micromech Microeng 15:10–14
24. Bhushan B (2004) Springer handbook of nanotechnology. Springer-Verlag, Berlin, Germany
25. Israelachvili JN (1998) Intermolecular and surface forces. Academic, London, UK
26. Majumder S, McGruer NE, Adams GG, Zavracky PM, Morrison RH (2001) Study of contacts in an electrostatically actuated microswitch. Sensor Actuat A Phys 93:19–26
27. Hah D, Yoon E, Hong S (2001) A low voltage actuated microelectromechanical switch for RF application. Jpn J Appl Phys 40(4B):2721–2724

Chapter 3
Dynamics of Elastic Vibro-Impact Microsystems

Abstract In this chapter the main ideas of the application of the finite element method for analysis of elastic microsystems are related to the theories of nonlinear vibro-impact oscillations, optimal control and design. The results obtained during simulations indicate that structural parameters of microsystem have significant influence on the dynamical process. Small variations of the parameters, which are common during MEMS fabrication, could be crucial for device performance. From another point of view, the presented herein new discovered phenomena of nonlinear mechanics are related to the possibilities to excite higher modes of vibrations of elastic links due to the local changes of cross-section, location and stiffness of contacting surfaces as well as by means of optimal configurations. This new approach helps designers to create stable and dynamically-reliable microstructures.

3.1 Mathematical Modeling

Figure 3.1 presents a schematic of the developed 2-D finite element (FE) model of impacting microcantilever, which represents the general case of vibro-impact microsystem (VIM). For simulation purposes a 2-D modeling approach is applied since: (a) flexural vibration modes have a much more significant influence on vibro-impact process in comparison to torsional modes and (b) it is computationally more cost-effective. The goal of the subsequent analysis is to focus on the impact process alone and carry out detailed investigation of important dynamic aspects of this complex phenomenon. Therefore in this chapter electrostatic forces are not considered and it is assumed that the microstructure is operating in vacuum, thus squeeze-film damping is neglected as well. Exclusion of gas environment from the presented FE model is justified by a preference to avoid ambient gas in device operation since it creates favorable conditions for electrical arching.

The model consists of $i = 1, 2, \ldots, m$ linear beam elements located in a single layer and $j = 1, 2, \ldots, k$ motion limiters or supports ($0 < k < 2m$) that are located in $i = 1, 2, \ldots, m$ nodes. Each beam element has two nodes with three degrees of

V. Ostasevicius and R. Dauksevicius, *Microsystems Dynamics*,
Intelligent Systems, Control and Automation: Science and Engineering 44,
DOI 10.1007/978-90-481-9701-9_3, © Springer Science+Business Media B.V. 2011

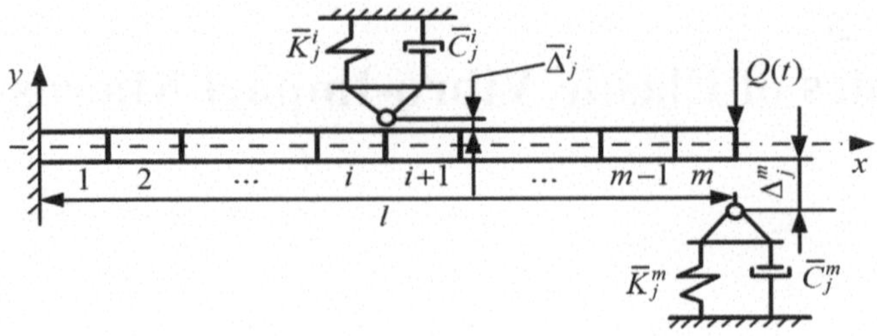

Fig. 3.1 Schematic of developed 2-D finite element model of impacting microcantilever

freedom (DOF) at each one (displacement in x- and y-axis directions and rotation in x_0y plane). The model was meshed manually with number of finite elements m equal to 50, thereby resulting in 150 total DOFs. The sufficiency of this particular mesh density was confirmed by comparative simulations presented in Section 3.2. Impact modeling is based on contact element approach and makes use of Kelvin-Voigt (viscoelastic) rheological model, in which linear spring is connected in parallel with a damper – the former represents the impact force and the latter accounts for energy dissipation during impact.

After proper selection of generalized displacements in the inertial system of coordinates, model dynamics is described by the following equation of motion given in a general matrix form:

$$[M]\{\ddot{y}(t)\} + [C]\{\dot{y}(t)\} + [K]\{y(t)\} = \begin{cases} \{Q(t)\}, if\{\bar{\Delta}^i_j\} > \{y_i(t)\} \cap \{\Delta^i_j\} < \{y_i(t)\} \cup f_i(y_i, \dot{y}_i, t) \geqslant 0; \\ \{Q(t)\} + \{F(y, \dot{y}, t)\}, if\{\bar{\Delta}^i_j\} \leqslant \{y_i(t)\} \cup \{\Delta^i_j\} \geqslant \{y_i(t)\} \cap f_i(y_i, \dot{y}_i, t) < 0, \end{cases}$$

where $[M]$, $[C]$, $[K]$ are mass, damping and stiffness matrices respectively, $\{y(t)\}$, $\{\dot{y}(t)\}$, $\{\ddot{y}(t)\}$ – displacement, velocity and acceleration vectors respectively. $\{Q(t)\}$ is a vector representing the sum of external forces acting on the microstructure. Since external electrostatic and air pressure forces are not considered here, this vector is used purely as a mechanical load during simulations.

The initial conditions are defined as:

$$\{\dot{y}(0)\} = \dot{y}^0; \{y(0)\} = \Delta^i_j,$$

where $\{F(y, \dot{y}, t)\}$ – vector of impact interaction between microcantilever and the support. Components $f_i(y_i, \dot{y}_i, t)$ represent the reaction of the impacting microstructure and are expressed as:

$$f_i(y_i, \dot{y}_i, t) = \bar{K}^i_j\left(\left|\Delta^i_j\right| - |y_i(t)|\right) + \bar{C}^i_j \dot{y}_i(t),$$

where $\bar{K}^i_j$, $\bar{C}^i_j$ – stiffness and viscous friction coefficients of the support, Δ^i_j— distance from the i-th nodal point of the microstructure to the j-th surface of the support located at the corresponding nodal point. In the case of the considered model the assumption of proportional damping is adequate therefore internal damping is modeled by means of Rayleigh damping approach [1]:

$$[C] = \alpha_{dM}[M] + \beta_{dK}[K],$$

where α_{dM}, β_{dK} are mass and stiffness damping parameters respectively that are determined from the following equations using two damping ratios ξ_1 and ξ_2 that correspond to two unequal natural frequencies of vibration ω_1 and ω_2 [1]:

$$\alpha + \beta\omega_1^2 = 2\omega_1\xi_1,$$
$$\alpha + \beta\omega_2^2 = 2\omega_2\xi_2.$$

The presented FE model was implemented in FORTRAN.

3.2 Free Transverse Impact Vibrations of Microcantilever

Free impact vibrations of elastic microstructures constitute one of the operation modes of contact-type MEMS devices. Complete vibro-impact process consists of free vibrations of the microstructure in the intervals between the impacts and its vibration during the impacts. Therefore, thorough analysis of free and impact vibrations of elastic microstructures is essential. For this purpose special FORTRAN numerical codes were written and used for running detailed dynamic simulations with the developed FE model of the microcantilever that undergoes impacts against the support.

The modes of natural transverse vibrations of elastic microcantilever-shaped links presented in Fig. 3.2 consist of transverse displacements Y (Fig. 3.2 (top)) and torsions Φ of the link components around the axes perpendicular to the plane of vibrations (Fig. 3.2 (bottom)). Of the whole range of natural vibrations, the first five modes were distinguished (I, II, III, IV, V) which in the intersection with the axis line formed nodal points marked by numbers that express the ratio between the distance x from the fixing site of the microcantilever and its whole length $l(x/l)$. As can be seen, with the increase of the number of the vibration mode, the number of nodal points increases.

The letters Y_n and Φ_n denote the values of the maximum amplitudes of displacement and torsional modes (maximum deflections). The vibration modes were obtained for a high degree of elastic link meshing of finite elements, $m = 100$. The process of free impact vibrations in the case when VIM have no prestress and their support is installed at the free end of the microcantilever is presented in Fig. 2.29 (top). This process is characterized by displacement z, speed, z' and P – the force of contact pressure between the support and the link.

The accuracy and adequacy of calculation results are greatly influenced by the extent of the finite elements mesh. Figure 3.3 presents the dependence of the maximum amplitude of the post-impact rebound $z_{\max} = y_{\max}/l$ on the position of the support at different density mesh. When $m = 50$, the obtained minimums of rebound amplitudes coincide with the nodal points of some vibration modes.

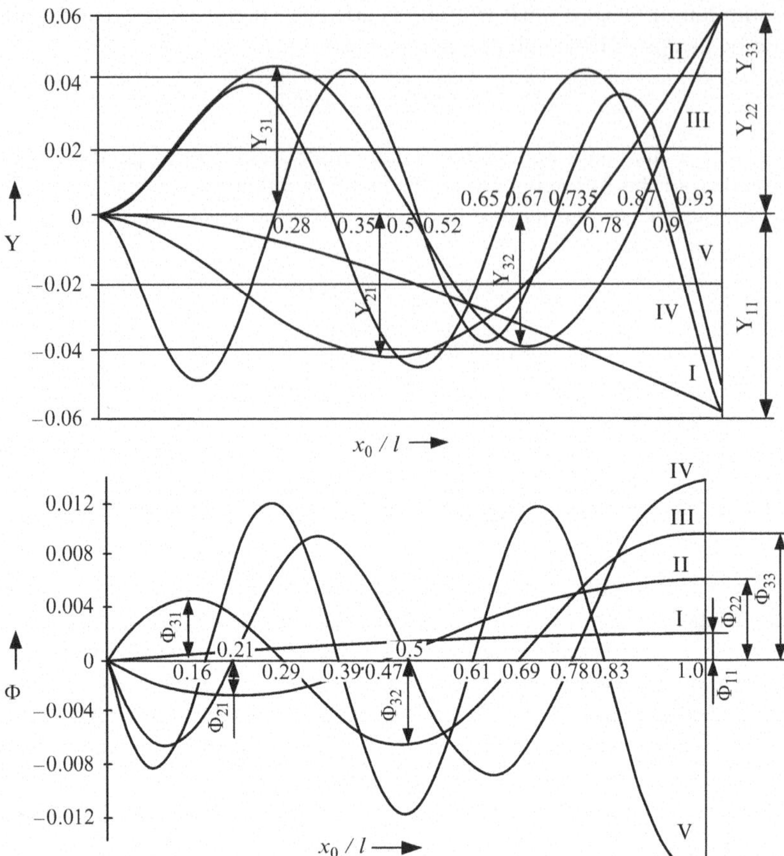

Fig. 3.2 Natural vibration modes of the microcantilever. *Top*: flexural (displacement) Y. *Bottom*: rotational (torsion) Φ. x_0/l denotes ratio between the distance x_0 from the anchor of the microcantilever and its whole length l, Y_{ij} and $Φ_{ij}$ – maximum amplitudes of the flexural and rotational modes respectively: index i – number of vibration mode, j – sequence number of the maximum amplitude point with respect to the anchor point

In the presence of a larger number of finite elements, it is possible to more accurately select the position of the support, which as a rule has to be positioned in the plane that separates two finite elements. In this case meshing of finite elements, $m = 50$, is sufficient and it more accurately reflects the dynamic processes of the VIM than meshing with $m = 10$ finite elements.

As can be seen from the diagram, the smallest rebound amplitudes are typical of such VIM in which the support is located in point $x_0/l = 0.87$ or $x_0/l = 0.67$. A slight decrease in the rebound amplitude is observed in point $x_0/l = 0.78$. The lower curve that asymptotically approaches the axis line corresponds to the deflection of the free end of the microcantilever during the impact on the support.

Figure 3.4 presents time characteristics of free impact vibrations in cases of meshing of finite elements, $m = 50$ and $m = 10$. When the support is installed in

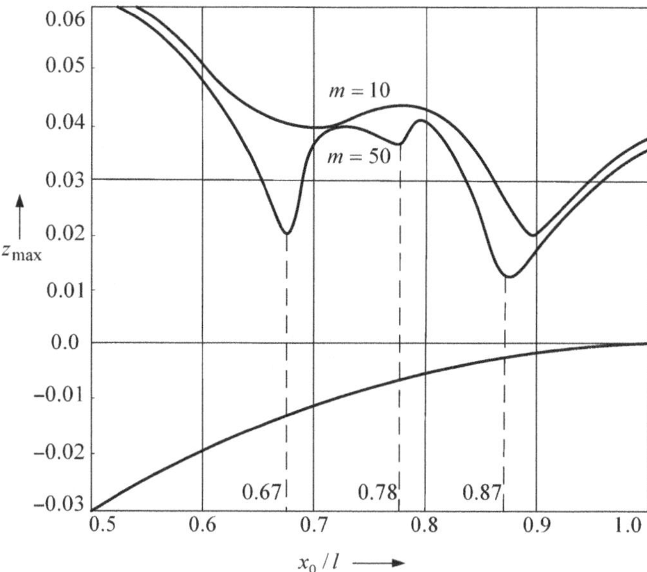

Fig. 3.3 Dependence of dimensionless rebound amplitude of the microcantilever $z_{max} = y_{max}/l$ on the position of the support expressed as a ratio between the distance x_0 from the anchor of the microcantilever and its whole length l

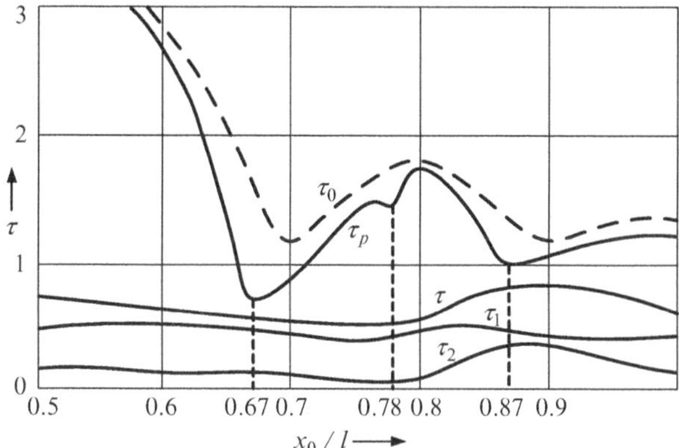

Fig. 3.4 Temporal characteristics of free impact vibrations of the microstructure as a function of support position: $\tau_p = \omega_1 T_p$ (for $m = 50$), $\tau_0 = \omega_1 T_p$ (for $m = 10$), $\tau = \omega_1 T$, $\tau_1 = \omega_1 T_1$, $\tau_2 = \omega_1 T_2$. ω_1 – first circular natural frequency of the microcantilever

points $x_0/l = 0.87, 0.78, 0.67$, the transient vibration time of the VIM $\tau = \omega_1 T_p$ may be shortened. The remaining time characteristics are less sensitive to changes in the position of the support though some of them can be increased.

These points are called particular points of natural modes. As the first three modes of transverse vibrations exert an important influence on the dynamic characteristics in the vibro-impact process, in design the choice may be confined to the

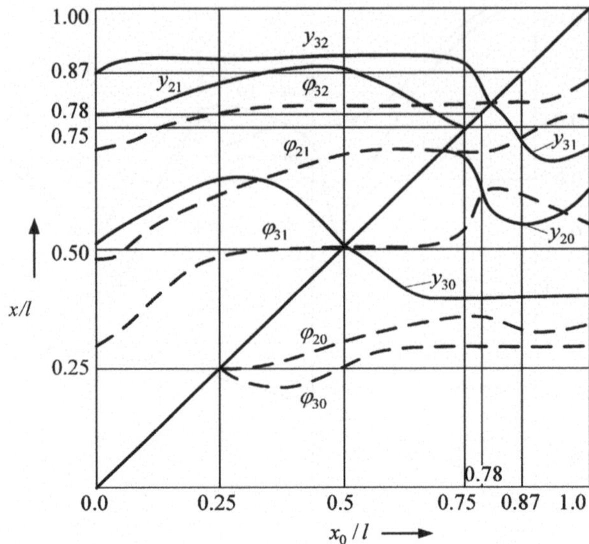

Fig. 3.5 Dependence of nodal points of the displacement (y_{ij}) and rotational (φ_{ij}) vibration modes of the supported microcantilever on the position of the support: i – number of vibration mode, j – sequence number of the nodal point with respect to the anchor point of the microstructure

three modes though the developed mathematical models account for a wider range of vibration modes.

To find the mechanical effects on dynamic characteristics, when the support is located in the particular points of modes, it is important to establish the modes of the link vibration during the impact on the support. For the purpose, the modes of the microcantilever-shaped structure with a support have been established, and dependences of the nodal points of the modes upon the location of support are presented in Fig. 3.5. The nodal points y_{ij} and φ_{ij} of the displacement mode Y_{ii} and torsional vibration mode Φ_{ii} are marked by two indices, the first of which meaning the number of vibration mode, and the second – the sequence number of the nodal point from the microcantilever fixing site.

Having introduced the support, one nodal point is added to each mode, as compared with the modes of the microcantilever free of support, marked as $i = 0$. In the diagram, the diagonal line represents the shifting of the support from the microcantilever fixing site to the free end. The position of the nodal points of vibration mode depends upon the kind of support. When the support is located at the microcantilever fixing site, the vibration modes and nodal points of the microcantilever coincide with the transverse vibration modes of non-supported microcantilever, which is demonstrated by the nodal points shown in the vertical axis *x*/*l*. Such location of nodal points characterizes the link before the impact on the support. When the support is shifted, part of the nodal points of the displacement mode are shifted together, though this is not typical of cross-section torsional modes. This phenomenon is related to the pin-joint support of the link.

We will now analyze some positions of the nodal points. For instance, nodal point y_{30} is displaced together with the support in the direction of the diagonal in Fig. 3.5 to the mid-point of the microcantilever, and point y_{20} is displaced still further whereas nodal point y_{31} joins the support at the moment when it reaches the mid-point of the microcantilever $(x_0/l = 0.5)$ and starts to move away when the support moves away from point $x_0/l = 0.78$.

The presented curves enable to explain the cause of changes in the dynamic characteristics of the VIM in the case when the support is located in point $x_0/l = 0.87$. In such position of the support, the second nodal point of the third vibration mode of the supported microcantilever coincides with the second nodal point of the third vibration mode of the same unsupported microcantilever ($x_0/l = 0.87$), which means that in the process of impact vibrations this point does not change its position either before or during the impact. When the force of impact is applied to the nodal point of the third mode of natural transverse vibrations of such mode are amplified. As the amplitude of third mode of vibrations becomes more intensive, the power dissipation in the material of the microcantilever increases since it is considered [1] that the energy dissipated by the microcantilever that vibrates in the higher mode exceeds that dissipated by the microcantilever that vibrates by the first one as many times as is the difference between the natural frequencies of the modes. Hence, the energy dissipated in the material of the microcantilever that vibrates in the first mode is by almost $\omega_3/\omega_1 \approx 17$ times less than the power dissipated in the material of the microcantilever that vibrates in the third mode. It is evident that the intensifying the amplitude of the third mode by locating the support at its nodal point does not terminate the first two modes of vibrations. The fact that nodal points y_{31} and φ_{21} coincide when the support is located at point $x_0/l = 0.87$ suggests the possibility of intensification of the second mode, yet anyway the advantages achieved in this case are first of all related to the intensifying the vibration mode amplitude of the microcantilever, namely.

The advantages achieved when the support is positioned in point $x_0/l = 0.78$ are related to the intensification of the second vibration mode amplitude because this is the point in which the nodal points of the second vibration mode of the supported and unsupported microcantilever are positioned ($x_0/l = 0.87$). Intersection of the nodal points y_{20} and φ_{31} displacement trajectories when the support is located in point $x_0/l = 0.78$, as presented in Fig. 3.5, also enables to intensify amplitude of other modes of transverse vibrations.

The above assertions are confirmed by the dependence of the maximal amplitude points of separate vibration modes upon the position of the support (Fig. 3.6). The relationship of amplitudes of displacement modes Y_{11} against the support locations (Fig. 3.6 (top)) shows that when the support is located in point $x_0/l = 0.87$, the amplitude of the third displacement mode Y_{33} is maximal whereas other amplitudes do not reach their maximal values in this point. Positioning of the support in the point of the maximum amplitude of the third vibration mode ($x_0/l = 0.67$) increases the displacement amplitude Y_{32} that coincides with the said point of maximum amplitude and amplitudes Y_{30} and Y_{31}, whereas amplitude Y_{33} decreases. When the support is positioned in the nodal point of the second displacement mode, the displacement

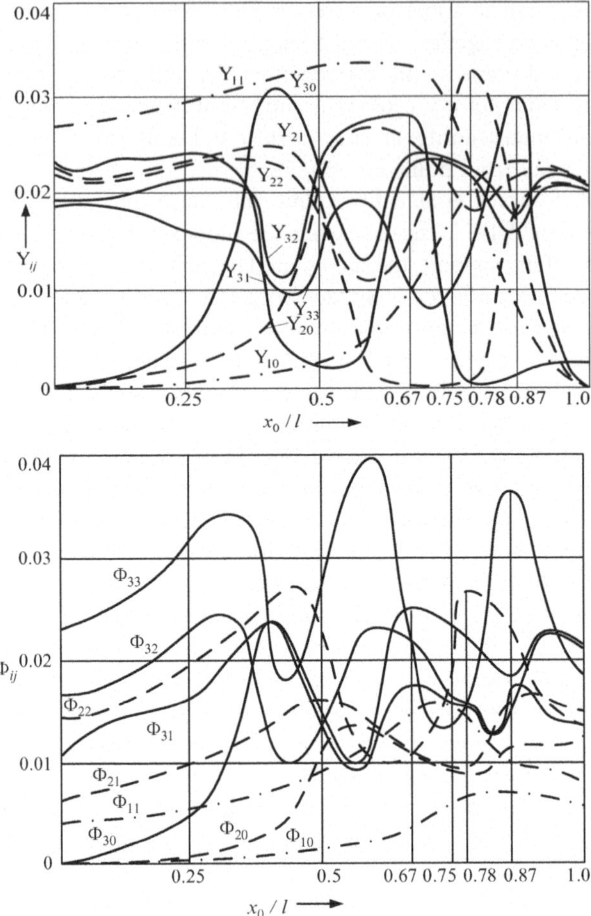

Fig. 3.6 Dependence of maximum amplitudes of the flexural (Y_{ij}) and rotational (Φ_{ij}) modes on the position of the support. *Top*: results for flexural modes. *Bottom*: results for rotational modes

amplitude Y_{22} increases whereas other amplitudes of the second mode decrease. Similarly, the amplitudes of torsional vibration modes Φ_{11} are increased, too (Fig. 3.6 (bottom)). Due to the impact of the microcantilever on the support located in one of the particular points of vibration modes, the said amplitudes increase still more, thereby intensifying separate vibration modes amplitudes.

After the analysis of the behavior of the nodal points and the points of maximum amplitude with regard to the support location along the axis of the microcantilever, it is important to investigate the dependence of the frequencies of separate vibration modes upon the position of the support. The dependences of the ratio between the natural frequencies of the supported microcantilever ω_i, and those of the unsupported microcantilever ω_{iin} are presented in Fig. 3.7. It may be seen that the first natural frequency I of the supported microcantilever reaches the maximum value

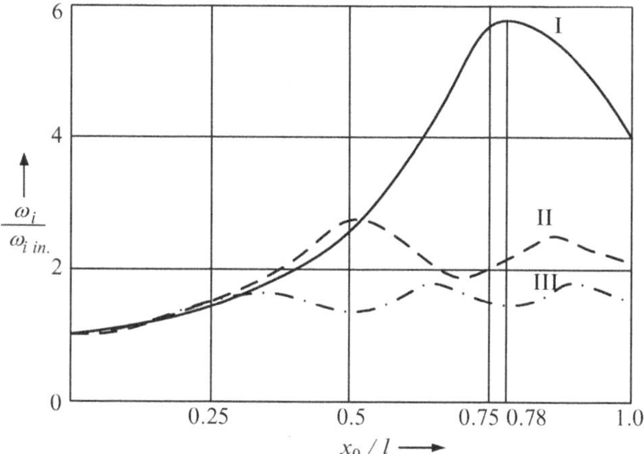

Fig. 3.7 Dependences of the ratio between the circular natural frequencies of the supported microstructure ω_i and those of the unsupported one ω_{iin} on the position of the support. The first natural frequency of the supported microcantilever I reaches the maximum value when the support is located in point $x_0/l = 0.78$ whereas the second II and the third III frequencies reach their maximum values when the support is located in other positions

when the support is located in point $x_0/l = 0.78$ whereas the second II and the third III natural frequencies reach their maximum values when the support is located in other positions. Therefore, in order to ensure maximum stability of the VIM containing a supported microcantilever, the support must be positioned in point $x_0/l = 0.78$. In such case the resonance frequency of the VIM is maximum, and, additionally, it becomes possible to amplify the amplitudes of the second mode of natural vibrations and to dissipate a marked portion of kinematically transferred energy in the material of the microcantilever. Furthermore, when the support is in point $x_0/l = 0.78$, the difference between the first and the second frequencies of the supported microcantilever is maximum, and by selecting the stiffness of the support to be located in the given point, the first natural frequency of the VIM may be brought closer to its second natural frequency, thereby increasing its vibrational stability.

The VIM may be manufactured with a gap between the impact-making links or with prestress. Therefore, it is important to know in what manner such gap or prestress is to be chosen for ensuring minimum link rebound amplitudes and minimum energy consumption in the VIM control.

Figure 3.8 presents maximum link rebound amplitudes $z_{max} = y_{max}/l$ depending upon the prestress Δ/l, when the support is located at the free end of the microcantilever. The diagonal line shows the position of the support when it is vertically moved from the boundary position to the position of maximum prestress. The dotted lines at the zero level represent the equilibrium position of the microcantilever (vertical) and zero prestress (horizontal).

As can be seen from the diagram, minimum amplitudes of the VIM rebound from the equilibrium position are characteristic in the presence of small prestresses (point B, when $\Delta/l = 0.01$). By drawing a perpendicular from point B to the

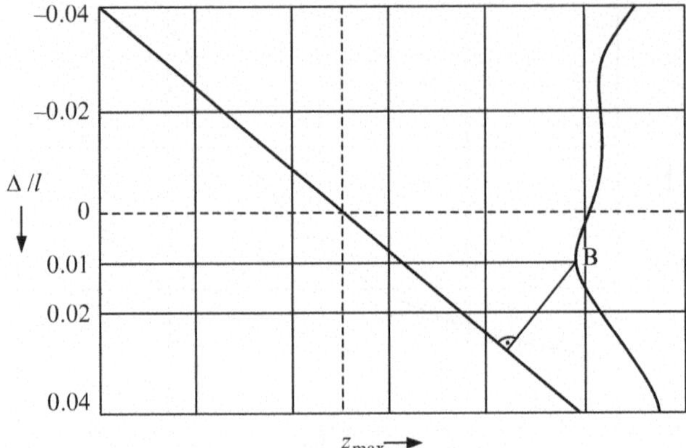

Fig. 3.8 Dependence of maximum rebound amplitudes of the microcantilever $z_{max} = y_{max}/l$ on the magnitude of dimensionless prestress Δ/l. Δ refers to prestress, i.e., displacement of the microcantilever free end in the direction perpendicular to the contact surfaces. Minimum rebound amplitudes are characteristic in the presence of small prestresses (point B, when $\Delta/l = 0.01$)

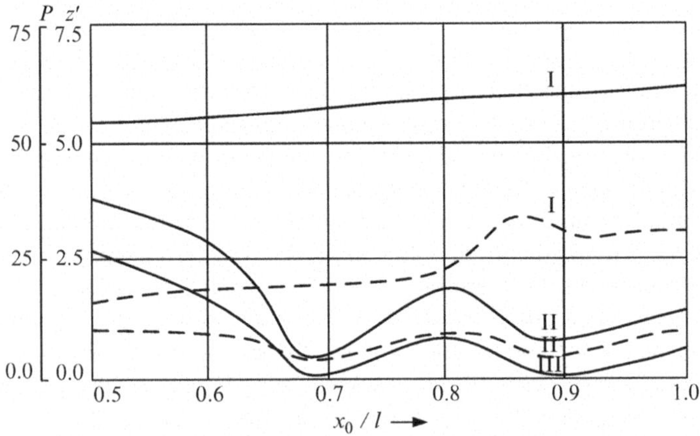

Fig. 3.9 Dependence of impact velocity z' (*continuous lines*) and contact pressure force P (*dashed lines*) on the position of the support. When the support is located in the particular points of the third transverse vibration mode of the microcantilever, a decrease in the velocity and original contact pressure force occurs, which is related to the increase in the dissipated energy in the link material

diagonal line, minimum amplitudes of the microcantilever rebound from the support are established. Thus, in order to obtain minimum rebounds of the VIM link with minimum power consumption, in the control of VIM the original prestress to be chosen should be established from point B.

In addition to the frequency characteristics of the amplitude of free impact vibrations, it is essential to determine the velocities and the forces arising during the link impact. Figure 3.9 demonstrates the dependence of the pre-impact velocity

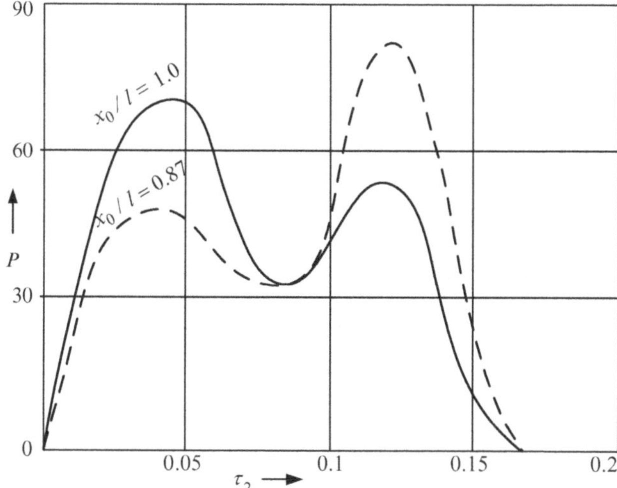

Fig. 3.10 Dependence of microcantilever impact force on the position of the support. During the microcantilever impact on the support positioned in point $x_0/l = 0.87$ the contact pressure force is lower in the first stage of impact than in the second one

(continuous lines) and original contact pressure force P (dashed line) upon the position of the rigid support, at zero prestress, during the first three microcantilever impacts (I, II, III) on the support. When the support is located in the particular points of the third transverse vibration mode of the microcantilever, a decrease in the velocity and original contact pressure force occurs, which is related to the increase in the dissipated energy in the link material.

Modeling has shown that during the microcantilever impact on the support positioned in point $x_0/l = 0.87$ the contact pressure force is lower in the first stage of impact than in the second one, as compared with the opposite characteristics of the pressure force when the support is located at the free end $x_0/l = 1.0$ (Fig. 3.10).

3.3 Temporal Characteristics of Vibro-Impact Microsystem with Rigid Support

A vibro-impact microsystem in which the microcantilever exerts impacts on a rigid support is investigated as a VIM with a rigid support. Such VIM may have a gap between the contact surfaces when these are in static equilibrium or have a prestress resulting from the displacement of the microcantilever by Δ distance in the direction that is perpendicular to the contact surfaces (Fig. 3.1).

First of all, it is necessary to determine the VIM variables expressed as time units. In this case the transient vibro-impact regime should be divided into three steps: the deformed position of the microcantilever 1 when its free end is lifted to height q from the contact surface of the rigid support 2; vibration period of the microcantilever 1 from the statically deformed position before the impact on the support 2; the period during the impact of the microcantilever 1 on the support 2.

To evaluate the static deformation of the microcantilever, we solve the equation:

$$EI\frac{\partial^4 y_1(x)}{\partial x^4} = 0,$$

where E – the elasticity modulus of the microcantilever material; I – inertia moment of the microcantilever cross-section.

Considering the boundary conditions

$$\left(\frac{\partial y_1(x)}{\partial x}\right)_{x=l_1} = \left(\frac{\partial^2 y_1(x)}{\partial x^2}\right)_{x=0} = 0,$$

when the origin of coordinates coincides with the contact point, we obtain the equation of the deformed microcantilever deflection:

$$y_1(x) = q + (q+\Delta)\left[\frac{1}{2}\left(\frac{x}{l_1}\right)^3 - \frac{3}{2}\left(\frac{x}{l_1}\right)\right]. \qquad (3.1)$$

This Eq. (3.1) describes the first step of the deformed state of the microcantilever.

Assuming that the microcantilever vibrates freely after its release till its impact against the rigid support, this motion law is described by the following equation:

$$EI\frac{\partial^4 y_1(x)}{\partial x^4} + \rho A\frac{\partial^2 y_1(x)}{\partial t^2} = 0, \qquad (3.2)$$

where ρA – mass of microcantilever length unit. This step corresponds to the boundary conditions:

$$\left(\frac{\partial y_1(x)}{\partial x}\right)_{x=l_1} = \left(\frac{\partial^2 y_1(x)}{\partial x^2}\right)_{x=0} = \left(\frac{\partial^3 y_1(x)}{\partial x^3}\right)_{x=0} = 0,$$

$$(y_1(x))_{x=l_1} = -\Delta,$$

where Δ – gap between the contact surfaces, and the initial conditions are:

$$y_1(x,0) = q + (q+\Delta)\left[\frac{1}{2}\left(\frac{x}{l_1}\right)^3 - \frac{3}{2}\left(\frac{x}{l_1}\right)\right], \frac{\partial y_1(x,0)}{\partial t} = 0.$$

For convenience, we express the solution of the Eq. (3.2) in the shape of superposition of natural modes:

$$f(x,t) = \sum_{i=1}^{\infty} f_i(x)\cos\omega_1 t, \tag{3.3}$$

where function

$$f(x,t) = y_1(x,t) + \Delta$$

has zero boundary and initial conditions:

$$f(x,0) = q + (q+\Delta)\left[\frac{1}{2}\left(\frac{x}{l_1}\right)^3 - \frac{3}{2}\left(\frac{x}{l_1}\right)\right], \frac{\partial f}{\partial t} = 0.$$

The frequencies of natural transverse vibrations of a microcantilever are calculated according to the equation:

$$\omega_i^2 = \frac{EI\mu_i}{\rho A l_1},$$

where μ_i – roots of the equation of frequencies.

For convenience, we write the solution of Eq. (3.3) as:

$$f(x,t) = (q+\Delta)\sum_{i=1}^{\infty} \varepsilon_i f_i\left(\frac{x}{l_1}\right)\cos\omega_1 t,$$

where

$$f_i\left(\frac{x}{l_1}\right) = (\cos\mu_i + ch\mu_i)\left(\sin\mu_i\frac{x}{l_1} + sh\mu_i\frac{x}{l_1}\right) - (\sin\mu_i + sh\mu_i) \times\left(\cos\mu_i\frac{x}{l_1} + ch\mu_i\frac{x}{l_1}\right),$$

$$\varepsilon_i = \frac{\int_0^{l_1} f\left(\frac{x}{l_1}\right)\left[1 + \frac{1}{2}\left(\frac{x}{l_1}\right)^3 - \frac{3}{2}\left(\frac{x}{l_1}\right)\right]dx}{\int_0^{l_1} f_i^2\left(\frac{x}{l_1}\right)dx}.$$

The value ε_i is a coefficient that defines the orthogonality of vibration modes.

We can easily calculate time T_1, from the start of the microcantilever motion till the impact:

$$y_1(l_1, T_1) = 0; (q + \Delta) \sum_{i=1}^{\infty} \varepsilon_i f_i(0) \cos \omega_i T_1 - \Delta = 0,$$

where

$$f_i(0) = -2(\sin \mu_i + sh\mu_i).$$

By denoting

$$\gamma = \Delta/(q + \Delta), \ \tau_1 = \omega_1 T_1$$

we obtain

$$-\sum_{i=1}^{\infty} 2(\sin \mu_i + sh\mu_i)\varepsilon_i \cos \frac{\omega_i}{\omega_1} \tau_1 = \gamma. \tag{3.4}$$

From the Eq. (3.4) we can easily determine time in the dimensionless form τ_1, of the microcantilever motion from the deformed position till the moment of the impact against the rigid support.

The third step, upon exerting an impact by its free end on the rigid support, the microcantilever continues moving in the same direction. As the motion of the free end of the microcantilever is constrained, in the time of impact we obtain a completely different diagram of the VIM calculation, consisting of the microcantilever with its free end pin-connected with the support. As the microcantilever is flexible and its variables are uniformly distributed along its full length, during the impact the medium part of the supported microcantilever vibrates. After some time the free end of the microcantilever rebounds off the rigid support. We will estimate the duration of the microcantilever impact on the support. The vibration of the microcantilever during the impact is described by the Eq. (3.2) for boundary conditions

$$(y_1)_{x=0} \left(\frac{\partial y_1(x)}{\partial x}\right)_{x=l_1} = \left(\frac{\partial^2 y_1(x)}{\partial x^2}\right)_{x=0} = 0, \ (y_1(x))_{x=l_1} = -\Delta$$

and initial conditions

$$y_1(x, 0) = \sum_{i=1}^{\infty} \varepsilon_i f_i \left(\frac{x}{l_1}\right) \cos \omega_i T,$$

$$\frac{y_1(x, 0)}{\partial t} = -\sum_{i=1}^{\infty} \varepsilon_i f_i \left(\frac{x}{l_1}\right) \omega_i \sin \omega_i T.$$

Substituting

$$y_1(x,t) = \Phi(x,t) + \Delta\left[\frac{1}{2}\left(\frac{x}{l_1}\right)^3 - \frac{3}{2}\left(\frac{x}{l_1}\right)\right],$$

we solve the equation

$$EI\frac{\partial^4\Phi(x,t)}{\partial x^4} + \rho A\frac{\partial^2\Phi(x,t)}{\partial t^2} = 0. \quad (3.5)$$

Taking into consideration the boundary conditions, we obtain:

$$\Phi(0,t) = \frac{\partial^2\Phi(0,t)}{\partial x^2} = 0,\ \Phi(l_1,t) = \frac{\partial\Phi(l_1,t)}{\partial x} = 0.$$

Considering the initial conditions

$$\Phi(x,0) = \sum_{i=1}^{\infty}\varepsilon_i f_i(x)\cos\omega_i T - \Delta\left[\frac{1}{2}\left(\frac{x}{l_1}\right)^3 - \frac{3}{2}\left(\frac{x}{l_1}\right)\right],$$

$$\frac{\partial\Phi(x,0)}{\partial t} = -\sum_{i=1}^{\infty}\varepsilon_i f_i(x)\omega_i\sin\omega_i T$$

we find the solution of equation

$$\Phi(x,t) = \sum_{m=1}^{\infty}\Phi_m(x)[\alpha_m\sin\omega_m t + \beta_m\cos\omega_m t],$$

where

$$\Phi_m(x) = sh\mu_m\sin\mu_m\frac{x}{l_1} - \sin\mu_m sh\mu_m\frac{x}{l_1},$$

μ_m – roots of the equation of frequencies of the microcantilever with a pin-connected support.

Coefficients α_m and β_m can be found in the same way as ε_i:

$$\alpha_m = \frac{\int_0^{l_1}\Phi_m(x)\left[\sum_{i=1}^{\infty}\varepsilon_i f_i(x)\omega_i\sin\omega_i T\right]dx}{\omega_m\int_0^{l_1}\Phi_m^2(x)dx},$$

$$\beta_{m=}\frac{\int_0^{l_1}\Phi_m(x)\left\{\sum_{i=1}^{\infty}\varepsilon_i f_i(x)\omega_i\cos\omega_i T-\Delta\left[\frac{1}{2}\left(\frac{x}{l_1}\right)^3-\frac{3}{2}\left(\frac{x}{l_1}\right)\right]\right\}dx}{\omega_m\int_0^{l_1}\Phi_m^2(x)dx}.$$

The time of the microcantilever impact against the support T_2, is found from the condition:

$$\left(\frac{\partial y_1}{\partial x^3}\right)_{x=0}=0.$$

By denoting $\tau_2=\omega_{m1}T_2$, we obtain:

$$\sum_{m=1}^{\infty}\mu_m^3\left[\alpha_m\sin\frac{\omega_n}{\omega_{m_1}}\tau_2+\beta_m\cos\frac{\omega_n}{\omega_{m_1}}\tau_2\right](sh\mu_m+\sin\mu_m)=0. \tag{3.6}$$

From the Eq. (3.6) it is easy to find τ_2 – time of the microcantilever contact with the rigid support in a dimensionless form.

Figure 3.11 provides the dependences of the duration of the microcantilever movement to the support $\tau_1=\omega_{m1}T_1$ (left) and the duration of the impact against the support $\tau_2=\omega_{m1}T_2$ (right) on the variables of the system.

3.4 Dynamic Characteristics of Microsystem Composed of Two Microcantilevers

The investigated VIM consists of two flexible microcantilevers (Fig. 3.12). Such VIM may have a gap between the links, may have no prestress and may have prestress between the microcantilevers 1 and 2. We will make calculations for a VIM with a prestress that arises upon displacing the microcantilever 1 along the axis y by distance Δ. Calculations are made for the transient motion law. The obtained results are also applicable for a VIM with a gap between the links.

As can be seen from Fig. 3.12 we may write:

$$l_1+l_2=L+\Delta l,$$

where Δl – the size of overlap is a negligible quantity as compared with the microcantilevers lengths l_1 and l_2 and the distance L between their constraint points.

Calculations for such VIM were carried out by using two methods: the method of energy balance that takes into account the main natural frequency, and the accurate method that takes into account the full range of natural frequencies of microcantilevers.

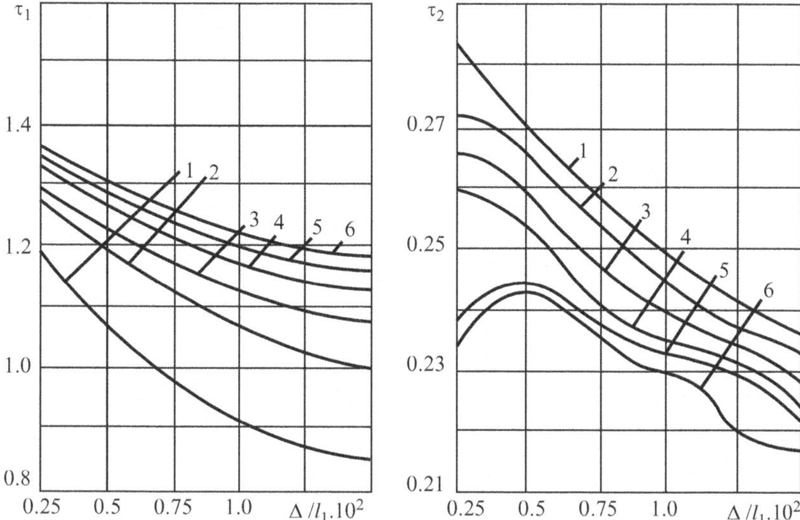

Fig. 3.11 Duration of microcantilever movement to the support (*left*) and impact against the support (*right*)

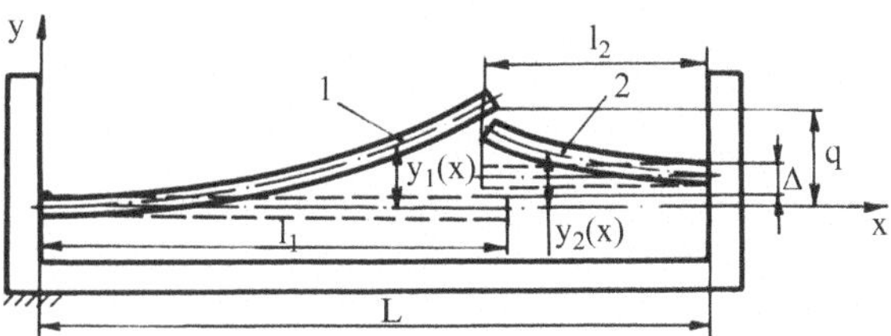

Fig. 3.12 Computational model of vibro-impact microsystem composed of two microcantilevers with lengths l_1 and l_2 and the distance L between their constraint points

In the case when the VIM has a static prestress, the microcantilevers, with their free ends contacting each other, deform each other and at the contact site create a certain initial contact pressure force.

When the microcantilevers of equal cross-sections are in mutual contact with each other, their deformed position, irrespective of the microcantilever thickness, may be described by the equation:

$$\frac{\partial^4 V_1(x)}{\partial x^4} = \frac{\partial^4 V_2(x)}{\partial x^4} = 0, \tag{3.7}$$

where $V_1(x)$, $V_2(x)$ – static deflections of the first and the second microcantilever.

When the microcantilevers are connected, we have the following boundary conditions:

$$V_1(x)|_{x=0} = 0,\ V_1(x)|_{x=L} = 0,\ \left.\frac{\partial V_1}{\partial x}\right|_{x=0} = \left.\frac{\partial V_2}{\partial x}\right|_{x=l} 0, \tag{3.8}$$

and the contact conditions may be expressed by the equations:

$$\begin{aligned}(V_1)_{x=l_1} &= (V_2)_{x=L-l_2},\\ \left(\frac{\partial^2 V_1}{\partial x^2}\right)_{x=l_1} &= \left(\frac{\partial^2 V_2}{\partial x^2}\right)_{x=L-l_2},\\ \left(\frac{\partial^3 V_1}{\partial x^3}\right)_{x=l_1} &= \left(\frac{\partial^3 V_2}{\partial x^3}\right)_{x=L-l_2},\end{aligned} \tag{3.9}$$

when the coordinates originate in the constraint point of the microcantilever 1, as shown in Fig. 3.12.

In many VIM the solution of the Eq. (3.7), with accurate evaluation of the Eqs. (3.8) and (3.9), acquires the following form:

$$V_1(x) = \frac{\Delta}{l_1^3 + l_2^3}\left(\frac{3}{2}l_1x^2 - \frac{1}{2}x^3\right),$$

$$V_2(x) = \frac{\Delta}{l_1^3 + l_2^3}\left(-\frac{3}{2}(L - x^2)l_2 + -\frac{1}{2}(l - x)^3\right).$$

Thus we obtain an expression that describes the deflection of any point of the microcantilever in the presence of a static deformation. It is not difficult to determine the static pressure force between the microcantilevers:

$$F_0 = \frac{3EI\Delta}{l_1^3 + l_2^3}$$

or in the dimensionless form

$$\eta_8 = \frac{3\eta_2\eta_7}{\eta_1{}^3 + 1}, \tag{3.10}$$

where $\omega_{1(2)}$ – the first natural frequency of the microcantilever 2.

$$\eta_1 = \frac{l_1}{l_2},\ \eta_2 = \frac{\Delta}{l_2},\ \eta_7 = \frac{EI}{\rho A l_2^2 \omega_{1(2)}^2},\ \omega_{1(2)} = 3.515\sqrt{\frac{EI}{\rho A l_2^4}}.$$

We shall determine the state of VIM when the free end of microcantilever 1 is released from the initial statically deformed position and under the effect of elastic straightening force makes impacts on the free end of the non-deformed microcantilever 2. The vibration of microcantilevers is described by the following known equations [2]:

$$\left.\begin{aligned} EI\frac{\partial^4 y_1}{\partial x^4} + \rho A\frac{\partial^2 y_1}{\partial t^2} = 0 \\ EI\frac{\partial^4 y_2}{\partial x^4} + \rho A\frac{\partial^2 y_2}{\partial t^2} = 0 \end{aligned}\right\}, \tag{3.11}$$

where y_1, y_2 – dynamic deflections of the first and the second microcantilevers during transverse vibrations.

Considering the boundary conditions

$$(y_1)_{x=0},\ (y_2)_{x=L} = \Delta,\ \left(\frac{\partial y_1}{\partial x}\right)_{x=0} = \left(\frac{\partial y_2}{\partial x}\right)_{x=L} = 0$$

and the contact conditions

$$\left.\begin{aligned} &(y_1)_{x=l_1} = (y_2)_{x=L-l_2} \\ &\left(\frac{\partial^2 y_1}{\partial x^2}\right)_{x=l_1} = \left(\frac{\partial^2 y_2}{\partial x^2}\right)_{x=L-l_2} = 0 \\ &\left(\frac{\partial^3 y_1}{\partial x^3}\right)_{x=l_1} - \left(\frac{\partial^3 y_2}{\partial x^3}\right)_{x=L-l_2} = \frac{\rho A}{EI}\frac{\partial^2 y_1}{\partial t^2} = P \end{aligned}\right\}, \tag{3.12}$$

where P – pressure force between contact surfaces.

We write the solution of Eq. (3.11) as:

$$\begin{aligned} y_1(x,t) &= V_1(x) + \alpha_1(x^3 - 3l_1x^2)\cos\omega t, \\ y_2(x,t) &= V_2(x) + \alpha_2\left[(L-x)^3 - 3l_1(L-x)^2\right]\cos\omega t, \end{aligned}$$

where ω – natural frequency of the microcantilevers at the moment of impact, α_1, α_2 – vibration amplitudes of the microcantilevers.

With account to the contact conditions of the microcantilevers, we find that the ratio of their vibration amplitudes:

$$a_1/a_2 = l_2^3/l_1^3.$$

In the application of the method of energy balance for the analysis of VIM, it is necessary to know the potential and the kinetic energy of the links.

The potential energy of links 1 and 2 is, respectively

$$\left.\begin{aligned} \Pi_1 &= \frac{EI}{2}\int_0^{l_1}\left\{\frac{\partial^2}{\partial x^2}\left[V_1(x) + \alpha_1(x^3 - 3l_1x^2)\cos\omega t\right]\right\}^2 dx \\ \Pi_2 &= \frac{EI}{2}\int_0^{l_2}\left\{\frac{\partial^2}{\partial x^2}\left[V_2(x) + \alpha_2(L-x)^3 - 3l_1(L-x)^2\right]\cos\omega t\right\}^2 dx \end{aligned}\right\}. \tag{3.13}$$

The kinetic energy of the links may be expressed by the equations:

$$\left.\begin{aligned} T_1 &= \frac{\rho A\omega^2}{2}\alpha_1\int_0^{l_1}\left(x^3-3l_1x^2\right)^2\sin^2\omega t dx; \\ T_2 &= \frac{\rho A\omega^2}{2}\alpha_2\int_0^{l_2}\left[(L-x)^3-3l_1(L-x)^2\right]\sin^2\omega t dx. \end{aligned}\right\}. \tag{3.14}$$

The energy balance condition for the investigated mechanical system in the transient motion law is:

$$\Pi_1+\Pi_2+T_1+T_2=\Pi_p, \tag{3.15}$$

where Π_p – potential energy of the deformed microcantilever 1:

$$\Pi_p=\frac{EI}{2}q^2\int_0^{l_1}\left\{\frac{\partial^2}{\partial x^2}\left[\frac{3}{2}\left(\frac{x}{l_1}\right)-\frac{1}{2}\left(\frac{x}{l_1}\right)^3\right]\right\}^2 dx.$$

By substituting (3.13) and (3.14) into (3.15), we obtain:

$$\begin{aligned} &\frac{EI}{2}\left[\int_0^{l_1}\left(\frac{\partial^2 V_1}{\partial x^2}\right)^2 dx+\int_0^{l_2}\left(\frac{\partial^2 V_2}{\partial x^2}\right)^2 dx\right] \\ &+18EI\alpha_1^2\cos^2\omega t\left[\int_0^{l_1}(x-l_1)^2dx+\left(\frac{l_1}{l_2}\right)^6\int_0^{l_2}(L-x-l_2)^2dx\right] \\ &+\frac{\rho A\alpha_1^2\sin^2\omega t}{2}\left\{\int_0^{l_1}\left(x^3-3l_1x^2\right)^2dx+\left(\frac{l_1}{l_2}\right)^6\int_0^{l_2}\left[(L-x)^3-3l_2(L-x)^2\right]^2dx\right\} \\ &=\frac{EI}{2}q^2\int_0^{l_1}\left\{\frac{\partial^2}{\partial x^2}\left[\frac{3}{2}\left(\frac{x}{l_1}\right)-\frac{1}{2}\left(\frac{x}{l_1}\right)^3\right]\right\}^2 dx. \end{aligned} \tag{3.16}$$

The Eq. (3.16) is correct in the case when the potential and kinetic energies of the structure are equal, i.e. provided that:

$$\begin{aligned} &18EI\int(x-l_1)^2dx+18EI\left(\frac{l_1}{l_2}\right)^6\int_0^{l_2}(L-x-l_2)^2dx \\ &=\frac{\rho A\omega_1^2}{2}\int_0^{l_1}\left(x^3-3l_1x^2\right)^2dx+\frac{\rho A\omega_1^2}{2}\left(\frac{l_1}{l_2}\right)^6\int_0^{l_2}\left[(L-x)^3-3l_2(L-x)^2\right]^2dx. \end{aligned}$$

From this equation we determine the main frequencies of the microcantilevers when they exert an impact against each other.

From the equality of potential energies

$$\frac{EI}{2}\left[\int_0^{l_1}\left(\frac{\partial^2 V_1}{\partial x^2}\right)^2 dx+\int_0^{l_2}\left(\frac{\partial^2 V_2}{\partial x^2}\right)^2 dx\right]$$
$$+18EI\alpha_1^2\left[\int_0^{l_1}(x-l_1)^2dx+\left(\frac{l_1}{l_2}\right)^6\int_0^{l_2}(L-x-l_2)^2dx\right] \tag{3.17}$$
$$=\frac{EI}{2}q^2\int_0^{l_1}\left\{\frac{\partial^2}{\partial x^2}\left[\frac{3}{2}\left(\frac{x}{l_1}\right)-\frac{1}{2}\left(\frac{x}{l_1}\right)^3\right]\right\}^2 dx$$

we can find the vibration amplitudes of the microcantilevers during the impact.

With the vibratory frequencies and amplitudes of the microcantilevers already known, we can easily determine the dynamic pressure force $F_d = -6EI\alpha_1\cos\omega t$ arising at the time of contact between the links:

$$F_d(t)>F_0.$$

By expressing it in a dimensionless form, we obtain

$$\eta_{10} = -6\eta_7\eta_9\cos\nu\tau,$$

where

$$\eta_9 = \alpha_1 l_2^2,\ \nu = \frac{\omega_1}{\omega_{1(2)}},\ \tau = \omega_{1(2)}t.$$

It is most important to determine the conditions of the microcantilever rebound from each other after the impact as this allows to judge about the motion law of the connected microcantilevers. It may be assumed that the contact disappears when the dynamic pressure force becomes smaller than the initial contact pressure force found from Eq. (3.10).

The dependences of the distance between the contact surfaces (η_2) on the ratio between the lengths of microcantilevers (η_1) in the presence of certain initial contact pressure forces (η_8) are provided in Fig. 3.13 (left). The vibration frequency ν of the microcantilevers whose dependence on the ratio between the lengths of the microcantilevers is provided in Fig. 3.13 (right), is minimal at the time of impact, when $\eta_1 = 1.415$. In this case the natural frequencies of transverse vibrations of the microcantilevers differ from each other by 2 times. As the expressions of kinetic energy (3.14) contain the square of vibratory frequency, it may be concluded that the kinetic energy of the structure, for the given length of microcantilevers, is minimal and for this reason the energy of higher modes increases. Therefore, it is probable that in the presence of prescribed lengths of microcantilevers, other characteristics of VIM have extreme values, too.

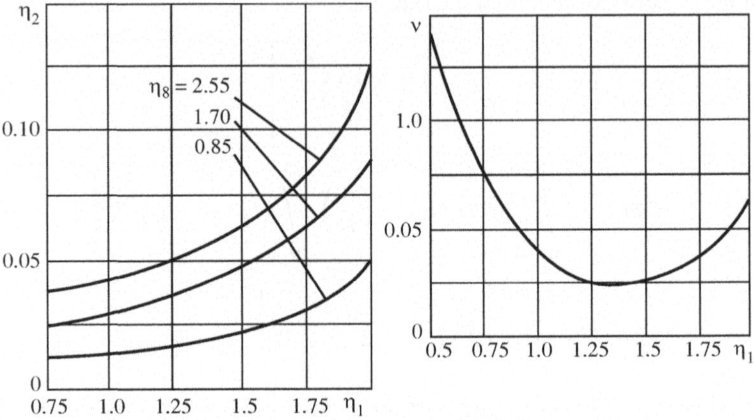

Fig. 3.13 Dependences of the initial contact pressure (*left*) and vibration frequency (*right*) on the ratio between the microcantilever lengths

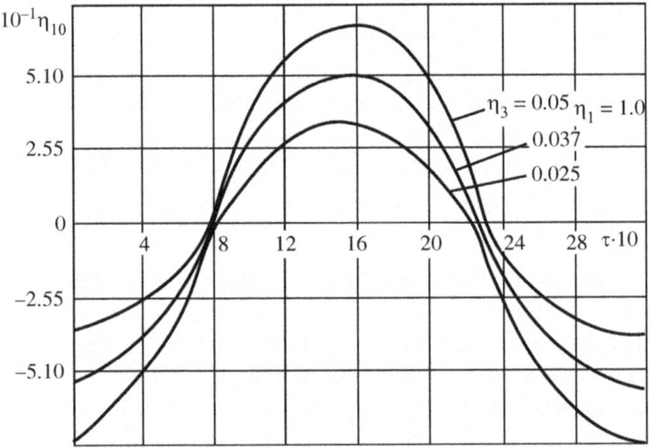

Fig. 3.14 Dynamic pressure force during the impact between the microcantilevers

Dynamic pressure force arising during the impact at the site of contact between the links depends on the deformation of microcantilever 1, $\eta_3 = q/l_2$ (Fig. 3.14).

For further improvement of the methods for VIM investigation, it is important to carry out profound studies of the impact-based interaction between flexible links by evaluating the complete range of vibration modes that have an effect on the dynamics of VIM. Also, it is important to assess the rheological properties of the contact surfaces.

When the free end of the microcantilever 1 is forced to height q, the following boundary conditions are applied:

$$(y_1)_{x=0}\left(\frac{\partial y_1}{\partial x}\right)_{x=0}=0,\ (y_1)_{x=l_1}=q,\ (y_1)_{x=0}\left(\frac{\partial^2 y_1}{\partial x^2}\right)_{x=l_1}=0.$$

By describing the deformed microcantilever by the cubic equation, we obtain:

$$y_1(x,0)=q\left[\frac{3}{2}\left(\frac{x}{l_1}\right)-\frac{1}{2}\left(\frac{x}{l_1}\right)^3\right].$$

After releasing the free end of the microcantilever 1, it moves towards the non-deformed link 2. The movement of the microcantilever 1 is described by the equation:

$$EI\frac{\partial^4 y_1}{\partial x^4}+\rho A\frac{\partial^2 y_1}{\partial t^2}=0. \tag{3.18}$$

Considering the boundary conditions

$$(y_1)_{x=0}\left(\frac{\partial y_1}{\partial x}\right)_{x=0}=0,\ \left(\frac{\partial^2 y_1}{\partial x^2}\right)_{x=l_1}=\left(\frac{\partial^3 y_1}{\partial x^3}\right)_{x=l_1}=0 \tag{3.19}$$

and the initial conditions

$$y_1(x,0)=q\left[\frac{3}{2}\left(\frac{x}{l_1}\right)-\frac{1}{2}\left(\frac{x}{l_1}\right)^3\right],\ \frac{\partial y_1(x,0)}{\partial t}=0 \tag{3.20}$$

the solution of the equation of free vibrations of the microcantilever 1 is found in the form of superposition of natural modes:

$$y_1(x,t)=\sum_{m=1}^{\infty}Y_m\left(\frac{x}{l_1}\right)(\sin\omega_m t+\cos\omega_m t),$$

where m – consecutive number of the free vibration mode of the microcantilever 1, $Y_m(x/l_1)$ and m-th function of the vibration mode described by the following equation:

$$Y_m\left(\frac{x}{l_1}\right)=C_1S\left(\mu_m\frac{x}{l_1}\right)+C_2T\left(\mu_m\frac{x}{l_1}\right)+C_3U\left(\mu_m\frac{x}{l_1}\right)+C_4V\left(\mu_m\frac{x}{l_1}\right),$$

where $\mu_m=k_m l_1$ – roots of the equation of natural frequencies of the microcantilever: C_1, C_2, C_3, C_4 – constants of integration.

With account to the boundary conditions (3.19) and by introducing the values of Duncan and Krylov function [2], the vibration modes of the microcantilever 1 are expressed by the equation:

$$Y_m\left(\frac{x}{l_1}\right)=\frac{C_3}{2}\left[\begin{array}{l}\left(ch\mu_m\frac{x}{l_1}-\cos\mu_m\frac{x}{l_1}\right)\\ -\frac{ch\mu_m+\cos\mu_m}{sh\mu_m+\sin\mu_m}\left(ch\mu_m\frac{x}{l_1}-\sin\mu_m\frac{x}{l_1}\right)\end{array}\right]. \quad (3.21)$$

For convenience, the solution of Eq. (3.18) is written as:

$$y_1(x,t)=q\sum_{m=1}^{\infty}Y_m\left(\frac{x}{l_1}\right)(A_m\cos\omega_m t+B_m\sin\omega_m t),$$

where A_m, B_m – coefficients found by using the orthogonality condition of the natural vibration modes of the microcantilever:

$$A_m=\frac{\int_0^{l_1}Y_m\left(\frac{x}{l_1}\right)\left[\frac{3}{2}\left(\frac{x}{l_1}\right)-\frac{1}{2}\left(\frac{x}{l_1}\right)^3\right]dx}{\int_0^{l_1}\left[Y_m\left(\frac{x}{l_1}\right)\right]^2dx},\ B_m=\frac{\int_0^{l_1}Y_m\left(\frac{x}{l_1}\right)\frac{\partial y_1(x,0)}{\partial t}dx}{\omega_m\int_0^{l_1}\left[Y_m\left(\frac{x}{l_1}\right)\right]^2dx}.$$

As derived from the initial conditions $B_m=0$.

Knowing that the scale of vibration modes is chosen freely, we shall introduce $C_3/2=1$ into the Eq. (3.21) and calculate the time T_1, of the movement of the microcantilever 1 from the initial deformed position till the impact on the microcantilever 2:

$$q\sum_{m=1}^{\infty}Y_m\left(\frac{x}{l_1}\right)A_m\cos\omega_m T_1=\Delta,$$

where $x=l_1$.

We can easily determine the velocity of the microcantilever 1 before the impact:

$$\frac{\partial y_1(x,t)}{\partial t}=-q\sum_{m=1}^{\infty}Y_m\left(\frac{x}{l_1}\right)\omega_m A_m\cos\omega_m T.$$

After the impact of the microcantilever 1 on the non-deformed microcantilever 2, both start to vibrate, and this can be expressed by the system of Eq. (3.11). The solution of the system (3.11) can be written as:

$$\begin{aligned}y_1(x,t)&=C_1S(k_nx)+C_2T(k_nx)+C_3U(k_nx)+C_4V(k_nx)',\\ y_2(x,t)&=C'_1S(k_nx)+C'_2T(k_nx)+C'_3U(k_nx)+C'_4V(k_nx)',\end{aligned}$$

where C'_1, C'_2, C'_3, C'_4 – constants of integration found from the boundary conditions for the microcantilever 2, $k_nL = \mu_n$ – unknown roots of the equation of natural frequencies of the connected microcantilevers.

The contact conditions of the microcantilevers are analogous to those described above by the (3.12) expressions where the equilibrium of the contact forces is evaluated:

$$EI\left(\frac{\partial^3 y_1}{\partial x^3}\right)_{x=l_1} = EI\left(\frac{\partial^3 y_2}{\partial x^3}\right)_{x=L-l_2} + P,$$

where P – support reaction in the contact zone of the microcantilevers.

The contact condition of the microcantilevers is satisfied if the deflection of the microcantilever 2 is expressed as:

$$y_2(x) = y_1(x) + \frac{P}{k_n^3 EI} V[k_n(x - l_1)]$$

and besides, the function $V[k_n(x - l_1)]$ is identically equal to zero when $x \leq l_1$, and not equal to zero when $x > l_1$.

Considering the boundary conditions of the constrained end of the microcantilever (1)

$$(y_1)_{x=0} = \left(\frac{\partial y_1}{\partial x}\right)_{x=0} = 0$$

we obtain $C_1 = C_2 = 0$ and the deflection of the microcantilever 1

$$y_1(x) = C_3 U(k_n x) + C_4 V(k_n x).$$

Then the expression of the deflection of the microcantilever 2 acquires the form:

$$y_2(x) = C_3 U(k_n x) + C_4 V(k_n x) + \frac{P}{k_n^3 EI} V[k_n(x - l_1)].$$

To find the natural frequencies of the vibrations of both connected microcantilevers, we will use the boundary conditions for the microcantilever 2, with no regard to the prestress of the VIM that has a small influence:

$$\left(\frac{\partial^2 y_1}{\partial x^2}\right)_{x=l_1} = (y_2)_{x=L} = \left(\frac{\partial y_2}{\partial x}\right)_{x=L} = 0$$

and thus we obtain a system of three equations for finding C_3, C_4 and $P/k_n^3 EI$:

$$\left.\begin{aligned}
&C_3U(k_nL)+C_4V(k_nL)+\frac{P}{k_n^3EI}V[k_n(L-l_1)]=0\\
&C_3T(k_nL)+C_4U(k_nL)+\frac{P}{k_n^3EI}U[k_n(L-l_1)]=0\\
&C_3S(k_nl_1)+C_4T(k_nl_1)=0
\end{aligned}\right\}. \quad (3.22)$$

By equating the determinant of the system (3.22) to 0, i.e.

$$\begin{vmatrix} U(k_nL) & V(k_nL) & V[k_n(L-l_1)] \\ T(k_nL) & U(k_nL) & U[k_n(L-l_1)] \\ S(k_nl_1) & T(k_nl_1) & 0 \end{vmatrix}=0,$$

we obtain the equation of vibration frequency of the connected microcantilevers:

$$\begin{aligned}
&T(k_nl_1)U(k_nL)U[k_n(L-l_1)]+U(k_nL)V[k_n(L-l_1)]\\
&=S(k_nl_1)V(k_nL)U[k_n(L-l_1)]+T(k_nl_1)T(k_nL)V[k_n(L-l_1)],
\end{aligned}$$

or

$$\begin{aligned}
&(chk_nl_1+\cos k_nl_1)(chk_nL-\cos k_nL)[shk_n(L-l_1)-\sin k_n(L-l_1)]\\
&+(shk_nl_1+\sin k_nl_1)(chk_nL-\cos k_nL)[chk_n(L-l_1)-\cos k_n(L-l_1)]\\
&=(chk_nl_1+\cos k_nl_1)(shk_nL-\sin k_nL)[chk_n(L-l_1)-\cos k_n(L-l_1)]\\
&+(shk_nl_1+\sin sk_nl_1)(shk_nL-\sin k_nL)[shk_n(L-l_1)-\sin k_n(L-l_1)].
\end{aligned} \quad (3.23)$$

In a partial case when $l_1=l_2$, by assuming that $k_nL=\mu_n$, we obtain:

$$\begin{aligned}
&(ch\frac{\mu_n}{2}+\cos\frac{\mu_n}{2})(ch\mu_n-\cos\mu_n)(sh\frac{\mu_n}{2}-\sin\frac{\mu_n}{2})+(sh\frac{\mu_n}{2}+\sin\frac{\mu_n}{2})(ch\mu_n-\cos\mu_n)(ch\frac{\mu_n}{2}-\cos\frac{\mu_n}{2})\\
&=(ch\frac{\mu_n}{2}+\cos\frac{\mu_n}{2})(sh\mu_n-\sin\mu_n)(ch\frac{\mu_n}{2}-\cos\frac{\mu_n}{2})+(sh\frac{\mu_n}{2}+\sin\frac{\mu_n}{2})(sh\mu_n-\sin\mu_n)(sh\frac{\mu_n}{2}-\sin\frac{\mu_n}{2}).
\end{aligned}$$

The following boundary conditions correspond to the connected microcantilevers:

$$(y_{1,2})_{x=0}=\left(\frac{\partial y_{1,2}}{\partial x}\right)_{x=0}=0,\ (y_{1,2})_L=\left(\frac{\partial y_{1,2}}{\partial x}\right)_{x=L}=0,$$

where $y_{1,2}$ – is the deflection of the connected microcantilevers. The equation of vibration modes acquires the form:

$$Y_n\left(\frac{x}{L}\right)=U\left(\mu_n\frac{X}{L}\right)-\frac{U(\mu_n)}{V(\mu_n)}V\left(\mu_n\frac{X}{L}\right),$$

or

$$Y_n\left(\frac{x}{L}\right)=\left(ch\mu_n\frac{x}{L}-\cos\mu_n\frac{x}{L}\right)-\frac{ch\mu_n-\cos\mu_n}{sh\mu_n-\sin\mu_n}\left(sh\mu_n\frac{x}{L}-\sin\mu_n\frac{x}{L}\right).$$

The trajectory of vibration of the connected microcantilevers is described as:

$$y_{1,2}(x,t) = \sum_{m=1}^{\infty} Y_m\left(\frac{x}{L}\right)(A_n \cos \omega_n t + B_n \sin \omega_n t), \tag{3.24}$$

where $\omega_n = \sqrt{\frac{EI}{\rho A}k_n^4}$ – vibration frequency of the connected microcantilevers, A_n, B_n – coefficients found from the condition of orthogonality of natural modes:

$$A_n = \frac{\int_0^L Y_n\left(\frac{x}{L}\right)dx}{\int_0^L \left[Y_n\left(\frac{x}{L}\right)\right]^2 dx} \int_0^{l_1} q \sum_{m=1}^{\infty} Y_m\left(\frac{x}{l_1}\right) A_m \cos \omega_m T dx;$$

$$B_n = -\frac{\int_0^L Y_n\left(\frac{x}{L}\right)dx}{\int_0^L \left[Y_n\left(\frac{x}{L}\right)\right]^2 dx} \int_0^{l_1} q \sum_{m=1}^{\infty} Y_m\left(\frac{x}{l_1}\right) A_m \omega_m \sin \omega_m T dx.$$

To determine the vibration amplitudes in the contact site, we introduce $x = l_1$ into the Eq. (3.24).

At the instant of the impact between the links, in the contact area there appear forces that deform the material of the contact surfaces. To determine the forces, it is convenient to employ Hertz theory on contact loads. For the purpose, we approximate the surfaces of the flat link in the contact area by paraboloid surfaces. We determine the coefficients of stiffness and damping that describe the properties of the contact surfaces.

In the general case the impact force is a non-linear time function, however, with account to the fact that it is big and gains its maximum value in a short time, it may be assumed that the pressure deformation of the contact surfaces and hence the impact force are linear time functions. The impact force is expressed as:

$$F_d = K_2 \sqrt{\alpha(t)^3}, \tag{3.25}$$

where $\alpha(t)$ – deformation of contact surfaces, K_2 – coefficient calculated according to the formula:

$$K_2 = \frac{4}{3\pi} \frac{1}{\delta_1 + \delta_2} \sqrt{\frac{R_1 R_2}{R_1 + R_2}},$$

where R_1, R_2 – radii of the contact surfaces, $\delta_1 = \delta_2 (1 - \nu^2)/(E\pi) = \delta_0$ – in the case when both microcantilevers are manufactured from the same material, ν – Poisson's ratio.

At the time instant when $dx/dt = 0$, the maximum pressure of the contact surfaces of the microcantilevers:

$$\alpha_m = \left\{ \frac{30\pi \left[q \sum_{m=1}^{\infty} Y_m \left(\frac{x}{l_1}\right) A_m \omega_m \sin \omega_m T \right]^2 \delta_0 m_{1pr} m_{2pr}}{16(m_{1pr} + m_{2pr})} \right\}^{\frac{2}{5}} \left(\frac{R_1 + R_2}{R_1 R_2} \right)^{\frac{1}{5}},$$

where m_{1pr}, m_{2pr} – reduced masses of the microcantilevers 1 and 2.

We assume that $\alpha(t)$ is a linear time function in the interval from $(\alpha)_{t=0} = 0$ to $(\alpha) = \alpha_m$ when t is equal to half of the impact time. We define the coefficient of stiffness of the contact surfaces. We have

$$F_{dmax} = C\alpha_m.$$

It may be derived that the stiffness coefficient of the contact surfaces $C = \frac{F_{d\max}}{\alpha_m}$ or $F_{d_{\max}} = K_2\sqrt{\alpha_m^3}$. Then we have $C = K_2\sqrt{\alpha_m^3}$ or $C = K_2\sqrt{\alpha_m}$.

From the presented formulae we can see that δ_0 and K_2 are constants but α_m and $F_{d\max}$, as well as the stiffnessies of the contact surfaces of the microcantilevers depend on the impact velocity, $\partial y_1(x,t)/\partial t$, however, in the presence of actual impact velocities the value of coefficient C alters but insignificantly and can be considered constant.

The dissipation of kinetic energy in the time of the impact between the microcantilevers must correspond to the amount of energy dissipated in the damper of viscous friction coefficient h. Determination of h is only possible by carrying out experimental tests of vibro-impact oscillations of the microcantilevers.

The ratio of two successive vibration amplitudes, when the connected microcantilevers vibrate defines the damped vibrations of the system and is permanently constant:

$$\frac{y_i}{y_{i+1}} = e^{\phi T}, \tag{3.26}$$

where $T = 1/\omega$ – period of vibrations, determined experimentally.

Decrease in the maximum deformation of the elastic link during two successive impacts is proportional to the decrease in the rebound height $y_i(x)$ of microcantilever 1 after two successive impacts.

We can write

$$\frac{y_i}{y_{i+1}} = \frac{y_m}{y_{m+1}}$$

or, by introducing the expression (3.26),

$$e^{\phi T} = \frac{y_m}{y_{m+1}}.$$

We obtain:

$$\phi = \frac{1}{T} \ln \frac{y_m}{y_{m+1}},$$

where quantity y_m expresses the deformation of microcantilever 1 at the moment of the impact.

Now we can easily calculate coefficient h:

$$h = \frac{\rho A l_1 \delta}{2T}, \tag{3.27}$$

where $\delta = \ln \frac{y_m}{y_{m+1}}$ – vibration decrement of the microcantilever 1, determined experimentally.

It is relevant to analyze the motion law of the microcantilevers during the impact as a vibration of preset amplitudes to which deformations of contact surfaces evaluated by rheological properties are added. The motion law of the contact surface of the microcantilever 1 with respect to the microcantilever 2 during the impact is expressed by the equation:

$$\frac{\partial^4 y}{\partial \xi^4} + \frac{\partial A l_1 \partial^2 y}{EI \partial t^2} - \frac{b l_1^3}{EI} \frac{\partial y}{\partial t} = \sum_{m=1}^{\infty} Y_m \left(\frac{x}{L}\right) (\bar{A}_n \cos \omega_n t + \bar{B}_n \sin \omega_n t), \tag{3.28}$$

where ξ – dimensionless coordinate, b – coefficient that determines the dissipation of energy, A_n, B_n – quantities calculated according to the formulae:

$$\bar{A}_n = A_n \left(1 - \frac{\mu_z^4}{\mu_n^4}\right), \; \bar{B}_n = B_n \left(1 - \frac{\mu_z^4}{\mu_n^4}\right),$$

where μ_z – roots of the equation of natural frequencies of the microcantilever leaning against the spring and the damper.

It has been experimentally established that almost half of the energy dissipated during the contact is attributed to the moment of impact. Keeping this in mind, we shall assume that in the studied VIM the energy is dissipated during the impact, and vibration decrement δ, determined during the vibro-impact motion law of the microcantilever 1, will be included in the formula (3.27) that expresses the viscous friction coefficient of contact surfaces.

It is possible to express the solution of the Eq. (3.28) in the form:

$$y(\xi, t) = \sum_{z=1}^{\infty} Y_z(\xi) T_z(t). \tag{3.29}$$

Then for the microcantilever we obtain:

$$\frac{\ddot{T}}{T} = -\frac{EI}{\rho A}\frac{\partial^4 y_1}{\partial \xi^4}\frac{1}{y}. \tag{3.30}$$

To have the Eq. (3.30) satisfied, both of its sides have to be constant. By denoting this constant as $-p^2$, we obtain two equations:

$$\left.\begin{array}{l} \dfrac{\partial^4 y}{\partial \xi^4} - \dfrac{\rho A p^2}{EI} y = 0 \\ \ddot{T} + p^2 T = 0 \end{array}\right\}. \tag{3.31}$$

The solution of the system of Eq. (3.31) can be expressed by the functions of Duncan and Krylov:

$$y(\xi) = C_1 S(\mu_z \xi) + C_2 T(\mu_z \xi) + C_3 U(\mu_z \xi) + C_4 V(\mu_z \xi).$$

For the boundary conditions

$$(y)_{\xi=0} = \left(\frac{\partial y}{\partial \xi}\right)_{\xi=0} = \left(\frac{\partial^2 y}{\partial \xi^2}\right)_{\xi=0} = 0,$$

$$EI\left(\frac{\partial^3 y}{\partial \xi^3}\right)_{\xi=1} = T(t) = Cy(1)T(t) + hy(1)\dot{T}(t),$$

we obtain the equation of natural frequencies of the microcantilever 1 leaning by its free end against the elastically viscous support:

$$\left[V(\mu_z)T(\mu_z) - S^2(\mu_z)\right]\frac{EI}{l_1}\mu_z^3 = (C + iph)[U(\mu_z)T(\mu_z) - S(\mu_z)V(\mu_z)]$$

$$(1 + ch\mu_z \cos\mu_z)\frac{EI}{l_1}\mu_z^3 + \left[C + i\frac{h\mu_z{}^3}{l_1^2}\sqrt{\frac{EI}{\rho A}}\right](ch\mu_z \sin\mu_z - \cos\mu_z \sin\mu_z = 0.$$

The roots of the equation of frequencies are obtained in the complex form:

$$\mu_z = a_z \pm b_z i.$$

It is easy to determine the vibration frequency

$$p_z = \omega_z \pm v_z i,$$

where $\omega_z = \frac{a_z^2 - b_z^2}{l_1^2}\sqrt{\frac{EI}{\rho A}}$ – circular frequency of vibrations of the microcantilever leaning against the elastically viscous support, $v_z = \frac{2a_z b_z}{l_1^2}\sqrt{\frac{EI}{\rho A}}$ – coefficient that evaluates the damping of vibrations $i = \sqrt{-1}$.

Time function *T*(*t*) is expressed by the equation:

$$T(t) = De^{(-v_z \pm i\omega_z)t} = De^{-v_z t}e^{\pm i\omega_z t},$$

where D – constant; $e^{-\omega_z t}$ – function represented by a sloping curve, and function $e^{\pm i\omega_z t}$ – by a sinusoid.

The expression of the natural vibration mode is obtained with consideration of the boundary conditions (3.19) as from (3.21):

$$Y_z(\xi) = \frac{C_3}{2}\left[(ch\mu_z\xi - \cos\mu_z\xi) - \frac{ch\mu_z + \cos\mu_z}{sh\mu_z - \sin\mu_z}(sh\mu_z\xi - \cos\mu_z\xi)\right].$$

It can be derived that the equation of natural vibration frequency can be written in the complex form:

$$Y_z(\xi) = \mathrm{Re}Y_z(\xi) + i\mathrm{Im}Y_z(\xi).$$

With regard to the fact that the scale of natural vibration frequency is freely selected, and knowing that

$$\mathrm{Im}Y_z(\xi) \ll \mathrm{Re}Y_z(\xi),$$

we may assume that

$$Y_z(\xi) = \mathrm{Re}Y_z(\xi). \tag{3.32}$$

Then by introducing the expression of natural modes into Eq. (3.32), we obtain:

$$Y_z(\xi) = cha_z\xi\cos b_z\xi - \cos a_z\xi chb_z\xi - sha_z\xi\cos b_z\xi + \sin a_z\xi chb_z\xi. \tag{3.33}$$

Considering (3.33), we introduce (3.31) into Eq. (3.29) and obtain the following result:

$$\sum_{z=1}^{\infty}\left[\ddot{T}_2(t) + p^2T_z(t)\right]Y_z(\xi) = \sum_{n=1}^{\infty}Y_n\left(\frac{l_1}{L}\right)(\bar{A}_n\cos\omega_n t + \bar{B}_n\sin\omega_n t). \tag{3.34}$$

By expanding the right side of the Eq. (3.34) by orthogonal eigenfunctions, we obtain the equation of the following form:

$$\ddot{T}_z(t) + p^2T_z(t) = \frac{\int_0^1 Y_z(\xi)d\xi}{\int_0^1 [Y_z(\xi)]^2 d\xi}\sum_{n=1}^{\infty}Y_n\left(\frac{l_1}{L}\right)(\bar{A}_n\cos\omega_n t + \bar{B}_n\sin\omega_n t).$$

The solution of the Eq. (3.34) can be written in the following form:

$$T_z(t) = \sum_{z=n=1}^{\infty} e^{-v_z t} \frac{\omega_n^2 \int_0^1 Y_z(\xi) d\xi}{\left(\omega_z^2 - \omega_n^2\right) \int_0^1 [Y_z(\xi)]^2 d\xi} Y_n\left(\frac{l_1}{L}\right)(\bar{A}_n \cos \omega_n t + \bar{B}_n \sin \omega_n t).$$

By introducing the result into (3.29), we obtain

$$y(\xi, t) = \sum_{z=n=1}^{\infty} e^{-v_z t} \frac{Y_z(1)\omega_n^2 \int_0^1 Y_z(\xi) d\xi}{\left(\omega_z^2 - \omega_n^2\right) \int_0^1 [Y_z(\xi)]^2 d\xi} Y_n\left(\frac{l_1}{L}\right)(\bar{A}_n \cos \omega_n t + \bar{B}_n \sin \omega_n t).$$

We obtain the relative movement trajectory of the contact surfaces of the microcantilevers during the impact.

We can easily determine the dynamic pressure force arising between the two microcantilevers at the moment of impact. From formula (3.25) we obtain

$$F_d = K_2 \sqrt{\alpha(t)^3},$$

where $\alpha(t) = y(\xi,t)$.

The maximum dynamic pressure force between the two microcantilevers appears at the time moment when:

$$\frac{d\alpha}{dt} = \frac{dy(\xi, t)}{dt} = 0.$$

From here we obtain

$$y(\xi, t) = \alpha_m.$$

Assuming that the contact in the VIM ends in the case when the dynamic pressure force becomes smaller than the prestress force, we can write the condition of vibrations in the contact site:

$$F_d(t) < F_0, \tag{3.35}$$

where F_0 – force of prestress.

This is how the post-impact disconnection of the microcantilevers can be determined.

The end of the vibration of the connected microcantilevers can be determined by using two conditions the first of which is expressed by formula (3.35), and the second one

$$y(\xi, t) \geqslant \delta$$

applies in the case when for microcantilever 1

$$y_1(x, T), \ \frac{\partial y_1(x, T)}{\partial t},$$

for microcantilever 2

$$y_2(x, T), \ \frac{\partial y_2(x, T)}{\partial t},$$

Introducing these conditions into the Eq. (3.21), we find the free vibrations of the microcantilevers. The second impact between the microcantilevers will occur in the case when

$$y_1(l_1, t) = y_2(L - l_2, t).$$

This impact is evaluated by using the above methods. We replace the initial conditions in the Eq. (3.24) that describes the vibration of the microcantilevers in the connected state by the coordinate of the impact between the microcantilevers and the velocity of their contact surfaces. In this manner calculations are made till the end of the vibro-impact movement of the microcantilevers. In the case when after the end of the movement non-impact vibrations of the microcantilevers continue, the state of the microcantilevers is described by using the method for the description of vibration of connected microcantilevers, which is different in that the value of the vibration damping decrement is determined in the stage of non-impact vibrations.

In the case when there is a gap between the contact surfaces of the VIM and the microcantilevers are not deformed, after the end of the vibro-impact movement a stage of free non-impact vibrations of each microcantilever follows. The characteristics of such vibrations are determined by using the methods for the description of pre-impact movement of such microcantilevers.

The above methods enable to describe the VIM consisting of two microcantilevers. By using the derived formulae, it is possible to calculate the main dependencies describing the VIM. The damping coefficient of contact surfaces is determined by introducing the microcantilevers vibration decrement into formula (3.27). The duration of the movement of the microcantilever 1 from the deformed position till the impact on the microcantilever 2 depends on the length of the microcantilever 1 (Fig. 3.15). By increasing the length of the microcantilever 1, we find that the duration of the pre-impact movement of the microcantilever 1 increases. The duration of the

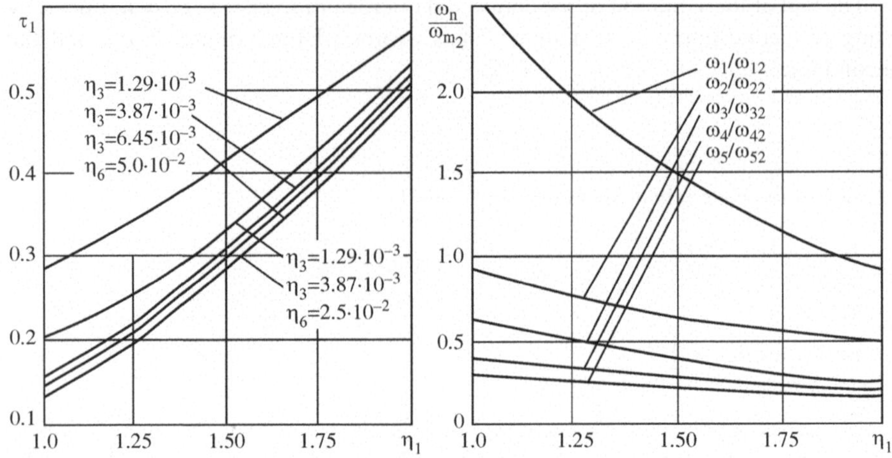

Fig. 3.15 Dependencies of the microcantilever 1 motion till the impact (*left*) and natural oscillation frequencies of the microcantilevers (*right*) upon the VIM parameters

pre-impact movement of microcantilever 1 decreases by increasing the prestress of the VIM or by decreasing the deformation degree of the microcantilever.

The complex manner of the alteration of the ratio between the vibration frequency of the connected microcantilevers ω_n and the dynamic pressure force of the natural frequencies of microcantilever 2 between the microcantilevers during the impact is also demonstrated by the characteristics of the dynamic pressure force that describes five vibration modes of the microcantilevers.

Notations in the diagrams are as follows:

$$\eta_4 = \frac{F_0 l_2^2}{EI}, \eta_5 = \frac{y_{1,2}}{l_2}, \eta_5 = \frac{F_d l_2^2}{EI}.$$

The dependencies of motion trajectories and contact pressure force during the impact between the microcantilevers are presented in Fig. 3.16.

3.5 Control of Vibration Modes of Links of Vibro-Impact Microsystems

Explanation of physical processes that occur during the microcantilever impact on the support enables to identify the possibilities of control of vibration modes and to improve the dynamic characteristics of VIM. As VIM operates in the regime of free impact vibrations, effective control of vibration modes is related to the choice of the VIM structure parameters. VIMs structure parameters may be changed by introducing additional masses in the contact areas, by bending straight links in different

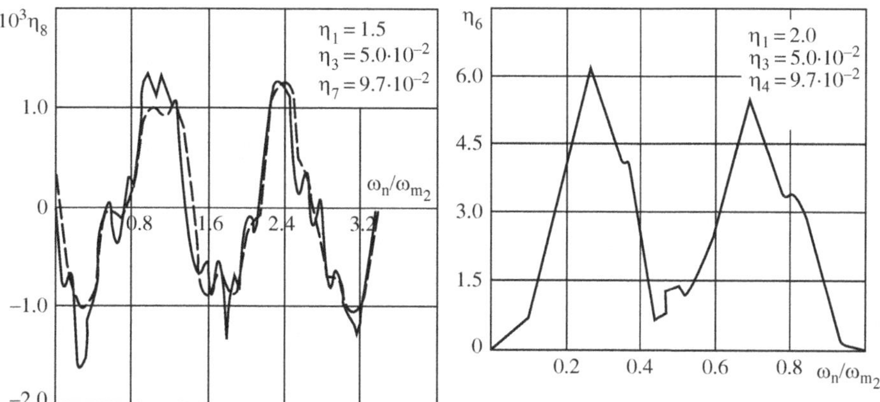

Fig. 3.16 Dependencies of motion trajectories (*left*) and contact pressure force (*right*) during the impact between the microcantilevers

places and at different angles, by changing the stiffness of the support, and by selecting the configuration of the elastic links. Namely such structural changes are actually possible in the mathematical models. After having established the effect of structural changes on the VIM dynamics, it is possible to create a structure that would ensure high quality of the VIM operation during exploitation.

By introducing additional concentrated "non-structural" masses to the elastic microcantilever, the natural frequencies may be changed, which evokes displacements of the nodal points of separate vibration modes. With account to the fact that the transient vibration law depends on the position of the support, introduction of additional masses allows to move the nodal point of a respective vibration mode to the area of the support. Such control of the positions of the points is convenient in the VIM in which the possibilities of moving the impact surfaces along the structure axis are limited. In accordance with the foregoing results of investigation, the vibro-impact process is mostly influenced by the particular points of the third and second modes of transverse vibrations. Therefore, we will show the possibility to control such modes.

Figure 3.17 (top) provides curves that represent the positions of the nodal point of the second vibration mode y_{21} and Fig. 3.17 (bottom) – the positions of the nodal point of the third mode y_{32} when the microcantilever is loaded with an additional mass. Horizontal dashed lines represent the positions of the nodal points of respective modes without additional masses, and the vertical ones – the beginning of the curves obtained for different ways of introducing additional masses.

The curve that starts at $x_m/l = 0.6$ is obtained after a fourfold increase of the mass of the microcantilever meshed by $m = 26$ finite elements, starting from $m = 16$ finite element. The curve that starts at point 0.64 was obtained after a twofold increase of the mass of two adjacent finite elements starting from finite elements $m = 16$ and $m = 17$, and the curve that starts in point 0.68 was obtained after a

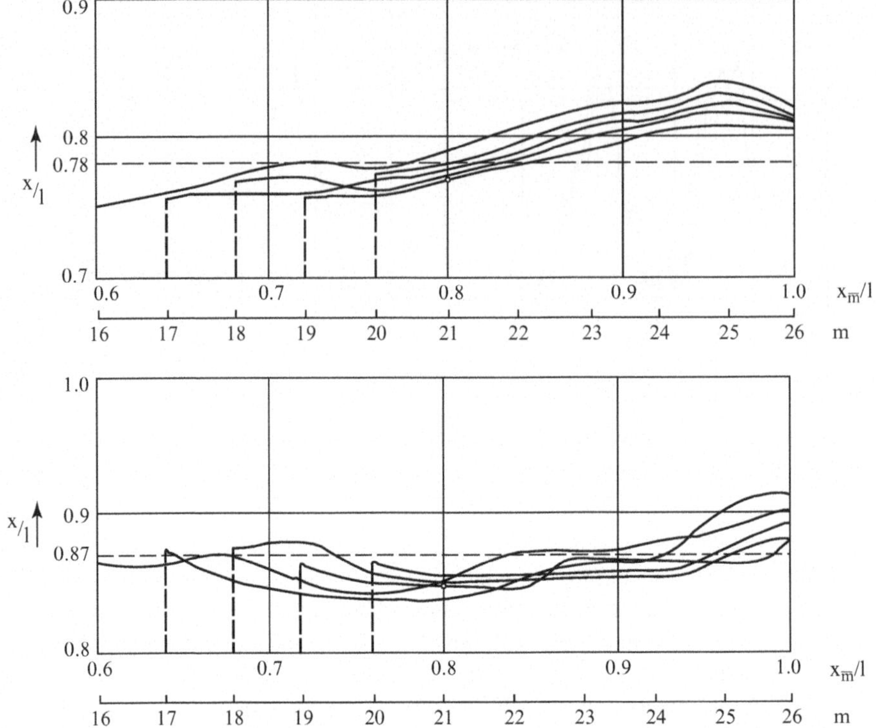

Fig. 3.17 Dependencies of the nodal points y_{21} and y_{32} of the second (*top*) and third (*bottom*) vibration modes on additional masses. Horizontal dashed lines represent the positions of the nodal points of respective modes without additional masses, and the vertical ones – the beginning of the curves obtained for different ways of introducing additional masses

twofold increase of the mass of each of the two finite elements with one finite element left in-between unchanged. The other curves were obtained after a twofold increase of the masses of finite elements with respectively two, three, etc. finite elements of mass left in-between unchanged.

Mounting of additional masses on the microcantilever has an effect both on the positions of the nodal points of the vibration modes and on the intensity of the amplitudes of such modes. The ratio between the amplitudes of vibration modes of a microcantilever with additional masses and respective amplitudes of the vibration modes of the original structure without additional masses is presented in Figs. 3.18 and 3.19, respectively for the second and the third modes of transverse vibrations. The analysis of the curves allows identifying the way of mass addition that enables to increase the amplitude of the relevant vibration mode, and it also reveals the merging tendency of several curves. Besides, the general decreasing tendency of vibration modes becomes evident with additional masses mounted at the free end of the microcantilever.

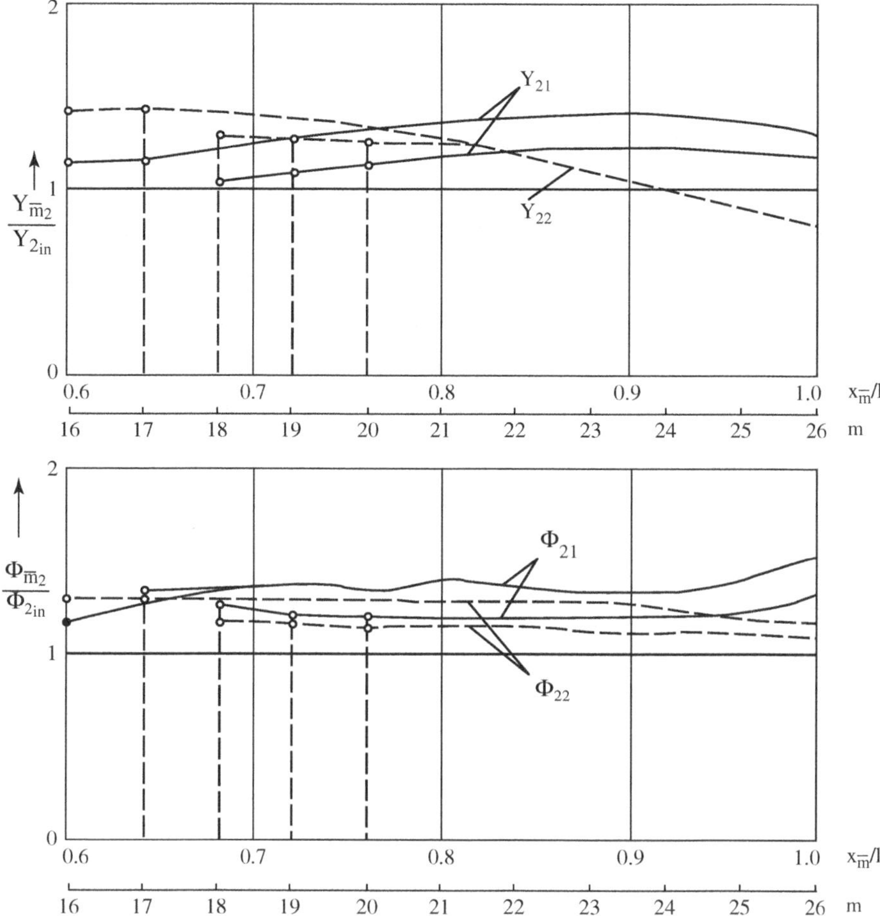

Fig. 3.18 Dependence of transverse displacement (*top*) and torsion (*bottom*) amplitudes of the second vibration mode upon additional masses. The general decreasing tendency of vibration modes becomes evident with additional masses mounted at the free end of the microcantilever

The frequency characteristics provide the most profound information about one or another way of introducing additional masses to the elastic structure. Figure 3.20 shows the dependences of the ratios between the frequencies of natural transverse vibrations of microcantilevers with added masses and the frequencies of elastic structures without additional masses upon the mass addition character. The worst results are obtained when additional masses are introduced in one cross-section of the structure, as suggested by the lowest values of frequency.

Another effective way to control the modes of elastic structure vibrations is by bending the elastic link.

To establish the effect of the link bending angle and site on the link frequency, Fig. 3.21 provides the dependences of the ratios of frequencies of bent and straight

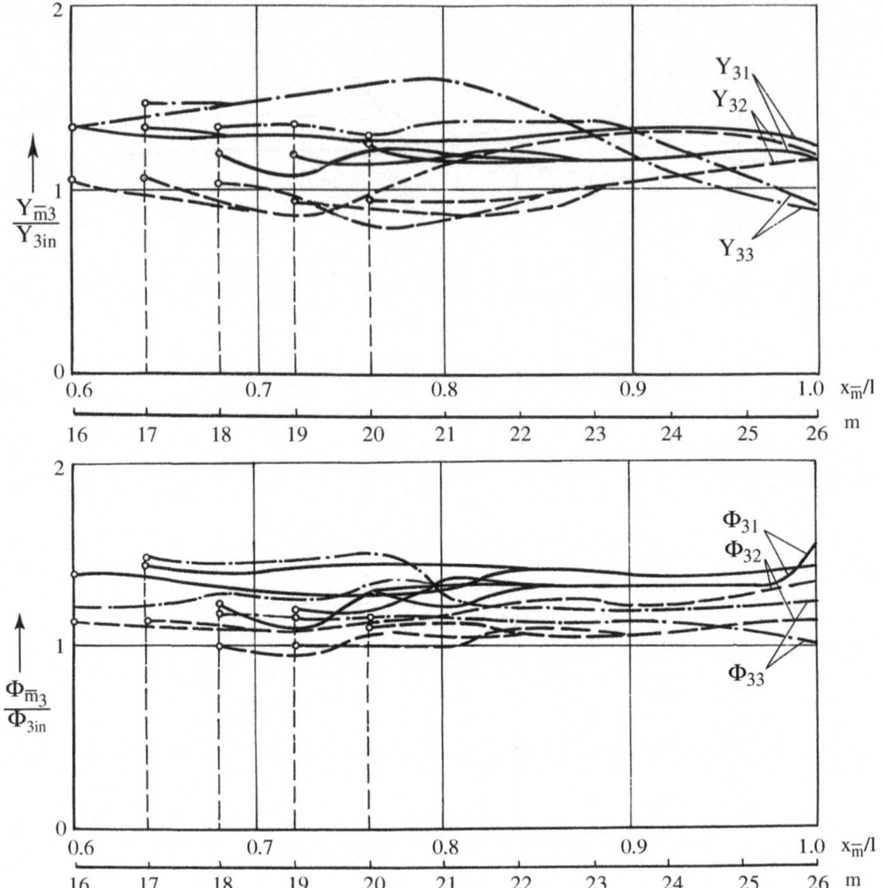

Fig. 3.19 Dependence of transverse displacement (*top*) and torsion (*bottom*) amplitudes of the third vibration mode upon additional masses

links upon the bending sites for different bending angles. As can be seen from the diagram, the lowest first natural frequency is characteristic of the microcantilever bent at the finite element $m = 18$, and the values of the second frequency are proximate when the structure is bent at $m = 14$ and $m = 12$ sites of finite elements. When $m = 0$, the values are obtained for a straight microcantilever. As can be seen from the curves, the first natural frequency of the bent structure is smaller than that of the straight one; therefore its vibrational stability is lower.

The modes of transverse vibrations may also be controlled by changing the stiffness of the support.

This applies to the control of vibration modes of a microcantilever that is in a free contact with the support. The aim of such control is to increase the vibrational stability of VIM by increasing the first frequency of transverse vibrations. Figure 3.22 provides dependences of the ratios between the natural frequencies of the supported structure

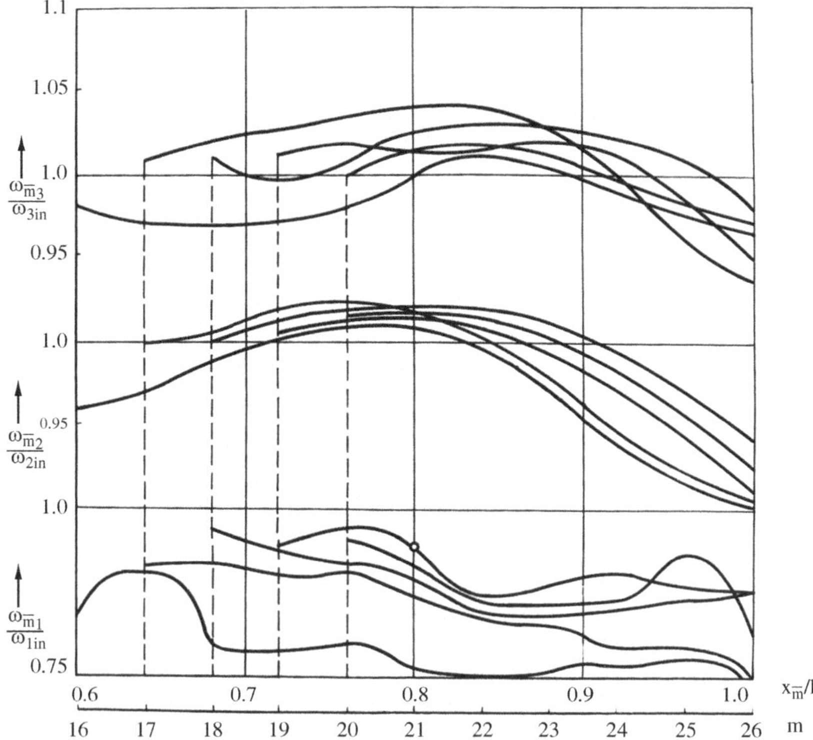

Fig. 3.20 Dependences of the natural frequencies of the microcantilever upon additional masses. The worst results are obtained when additional masses are introduced in one cross-section of the structure, as suggested by the lowest values of frequency

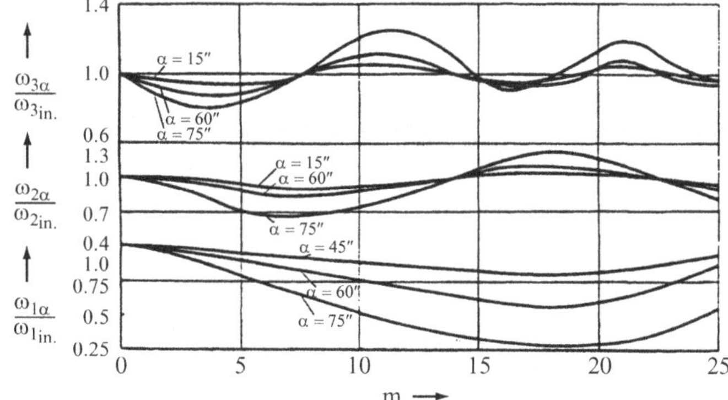

Fig. 3.21 Dependence of natural frequencies of the microcantilever upon the size and place of the bending angle. The first natural frequency of the bent structure is smaller than that of the straight one therefore its vibrational stability is lower

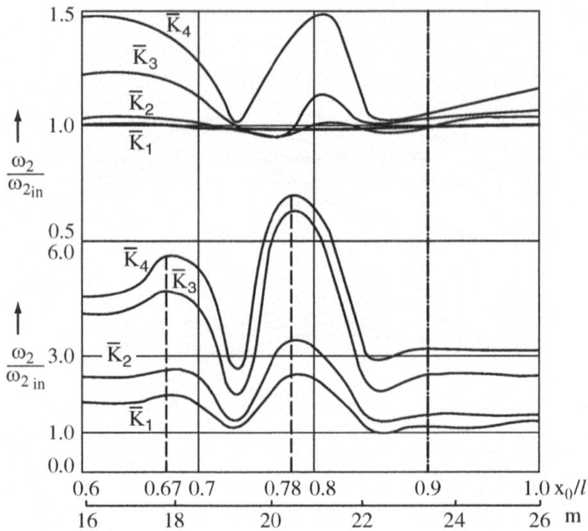

Fig. 3.22 Dependence of the natural frequencies of the supported microcantilever upon the place and stiffness of the support, when the support is in this case at the nodal point of the second vibration mode $x_0/l = 0.78$, it is possible to markedly increase the first frequency of VIM and to bring it closer to the second frequency

and the respective frequencies of the unsupported structure, where $i = 1{,}2$ – vibration mode number, upon the support stiffness and location, which show that as the stiffness of the support increases $\bar{K}_4 > \bar{K}_3 > \bar{K}_2 > \bar{K}_1$ and reaches a certain value $\bar{K}_3$ or $\bar{K}_4$, if the support is in this case at the nodal point of the second vibration mode $x_0/l = 0.78$, it is possible to markedly increase the first frequency of VIM and to bring it closer to the second frequency.

The analysis of the VIM vibration modes when the stiffness of the support is $\bar{K}_3$ showed that the first vibration mode completely disappears. This is confirmed by the appearance of the nodal points y_{11} and φ_{11} in the first mode of transverse vibrations which is not characteristic of the investigated VIM with an elastic support (Fig. 3.23). An increase in the first natural frequency and a respective appearance of nodal points are also observed when a support of equal stiffness is located in the maximum amplitude point of the third vibration mode $x_0/l = 0.67$. It is noteworthy that when the support is located at this point, the nodal points that appeared in the first vibration mode almost coincide with the first particular points of the third mode y_{31} and φ_{31}. If the support is in the point $x_0/l = 0.78$, the nodal points appear closer to the free end of the structure. The nodal points of the second mode abruptly change their position when the support of stiffness $\bar{K}_3$ is close to the point $x_0/l = 0.78$.

The stiffness of the support, $\bar{K}_3$ corresponds to such value in the presence of which the natural frequency of the support equals or exceeds the second frequency of transverse vibrations of the unsupported microcantilever. To make the natural frequency of the support equal to the second natural frequency of the elastic microcantilever, the natural frequency of the microcantilever spring-shaped support

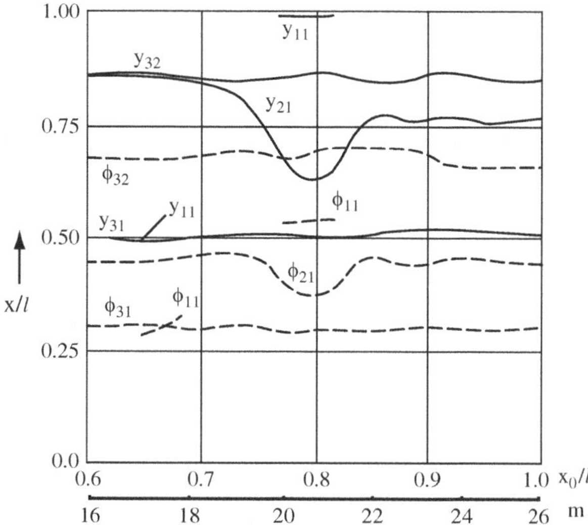

Fig. 3.23 Dependence of the position of the nodal points of oscillation modes of the supported microcantilever upon the position of the support, when the support rigidity is $\bar{K}_3$ the first vibration mode completely disappears. This is confirmed by the appearance of the nodal points y_{11} and φ_{11} in the first mode of transverse vibrations which is not characteristic of the investigated VIM with an elastic support

should be by 2.51 times smaller than the length of the microcantilever, the other structural parameters of the VIM being identical. Then the stiffness of the support by 16.5 times exceeds the stiffness of the microcantilever.

The VIM with an elastic – support of $\bar{K}_4 > \bar{K}_3$ stiffness has extended areas of nodal points y_{11} and φ_{11} close to the particular points $x_0/l = 0.67$ and 0.87, and the nodal points y_{21} and φ_{21} approach zero when the support is located in point $x_0/l =$ 0.78. Besides, when the support is located at the free end of the microcantilever, nodal point φ_{11} appears.

As a result of the analysis, it was found that with the help of an elastic support, the second or the third mode of transverse vibrations can be excited and thereby the vibrational stability of VIM is enhanced.

VIM modeling program not only enables to choose the above described structure parameters but also to set any configuration for the elastic structure in the presence of any boundary conditions. The results of modeling were as follows.

The dependence of maximum rebound amplitudes $z_{max} = y_{max}/l$ upon the position of the support for zero prestress (Fig. 3.24) proves that the configuration of the link has an effect on its vibration modes. Minimum rebound amplitudes in the presence of a support located in different points of the microcantilever prove the effect of the structure configurations on the positions of the nodal points of vibration modes, and the decrease in the intensity of rebound amplitudes suggests the possibility to intensify amplitudes of one of the vibration modes by selecting a relevant configuration of the structure. In this aspect, the most characteristic is the dot and dash line

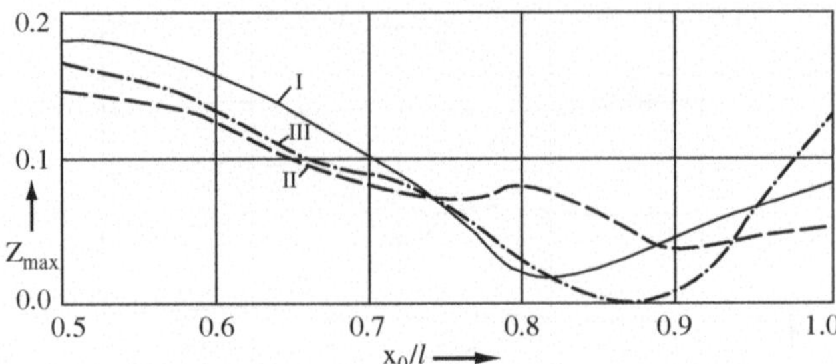

Fig. 3.24 Dependence of rebound amplitudes of the optimal microcantilever upon the position of the support. Minimum rebound amplitudes in the presence of a support located in different points of the microcantilever prove the effect of the structure configurations on the positions of the nodal points of vibration modes, and the decrease in the intensity of rebound amplitudes suggests the possibility to intensify amplitudes of one of the vibration modes by selecting a relevant configuration of the structure

obtained for an optimal structure for the prescribed third frequency of natural transverse vibrations (Fig. 3.49). A marked decrease in the rebound amplitudes when the support is located in point $x_0/l = 0.87$ shows that such interaction of the VIM impacts results in the excitation of the third vibration mode only. As it will be shown later on, the third vibration mode is excited due to the fact that the configuration of the elastic structure is optimal for the prescribed third frequency, and in its excitation in the wide range of frequencies, as is the case during the impact, the elastic link acts as a resonator (the dashed line represents the optimal configuration of the second form (Fig 3.50), while continuous line – of the first form (Fig. 3.49)).

A marked effect of the structure configuration on the contact pressure force $P = F/F_0$, where F_0 is the prestress, occurs when the support is at the free end of the microcantilever (Fig. 3.25). Dot-dash line obtained for the prescribed third frequency optimal structure coincides with the character of dashed line on Fig. 3.10 for the support position in point $x_0/l = 0.87$ of the nodal point of third natural frequency.

3.6 Analysis of Forced Vibrations of Microcantilever

The behavior of VIM in the case of harmonic excitation is best defined by the areas of existence of diverse vibration laws. The most interesting in the practical aspect is the settled law when during one excitation period the microcantilever makes one impact on the rigid support. It is clear that the width and location of the settled laws depends on the variables and the excitation type of the VIM structure.

Figure 3.26 provides areas of the existence of diverse vibration impact laws in the case of sine excitation when the support is at the free end of the microcantilever. When the VIM gap $\Delta = 0$, the excitation amplitude is in all cases the same. As can be seen from the diagram, in this case the area of settled periodic law (right-hand dashes) is the widest. When the frequency ω of the excitation force is increased, in

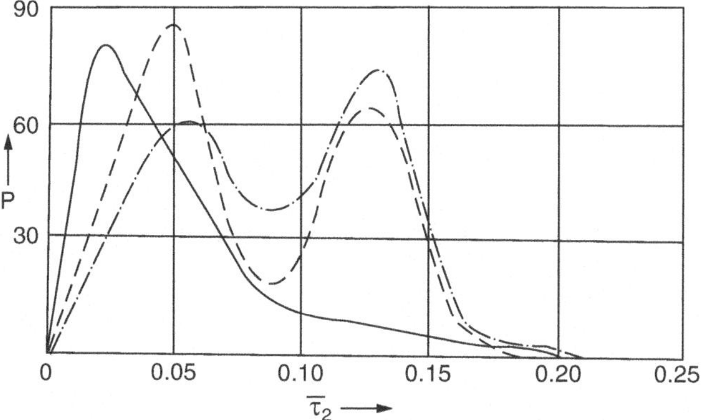

Fig. 3.25 Effect of the microcantilever shape upon the character of the impact force

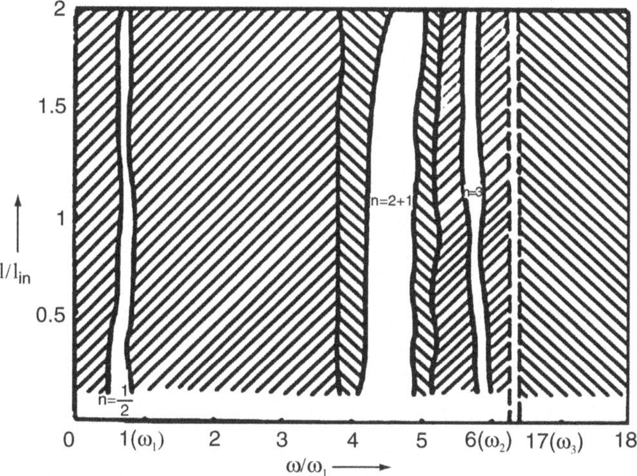

Fig. 3.26 Areas of diverse vibration impact laws at sine excitation: settled periodic law (*right-hand dashes*), unsettled vibration law (*left-hand* dashes) and n-recurrent motion law

the case when it becomes by four times higher than the first natural frequency $\omega_1(\omega/\omega_1 = 4)$, the unsettled vibration law appears (left-hand dashes), irrespective of the ratio between the microcantilever length l and the initial structure length $l_{in}(l/l_{in})$. At further increase of the excitation frequency recurrent laws of settled vibration appear. Law $n = 2 + 1$ means that in the course of two excitation periods the microcantilever makes one impact on the support and subsequently one impact in one excitation period. Law $n = 3$ means that in the course of three excitation periods the microcantilever makes one impact, and law $n = 1/2$ – means that in the course of one excitation period the microcantilever makes two impacts. Such law is characteristic of elastic VIM and is achieved upon excitation of the spectrum of transverse vibration modes.

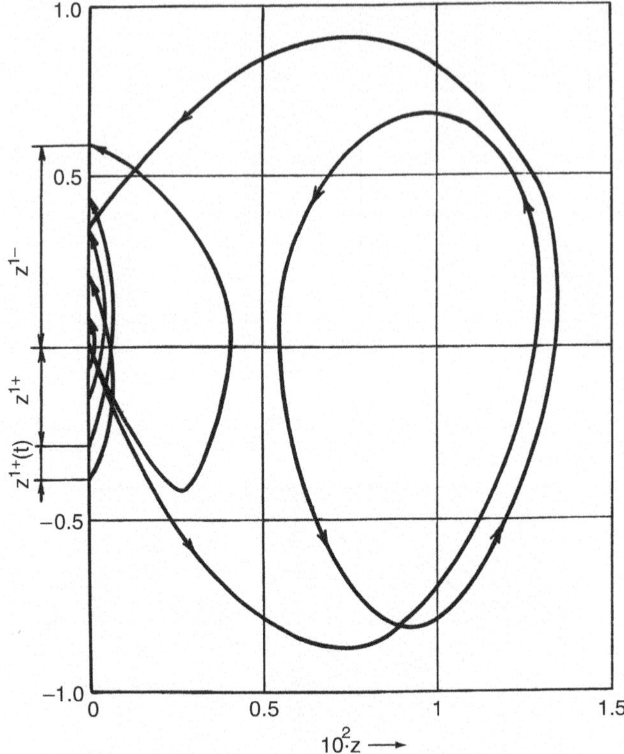

Fig. 3.27 Phase trajectory of vibro-impact n = 2 + 1 recurrent motion law of the microcantilever

For a better explanation of law $n = 2 + 1$ Fig. 3.27 provides the phase trajectory of vibro-impact motion law of the free end of the microcantilever when the excitation frequency is $\omega/\omega_1 = 4.5$. In the horizontal axis displacement $z = y/l$ is presented whereas in the vertical axis the pre-impact velocity $z^{1-} = \dot{y}^{-}\rho A\omega_1 l/F_{in}$and post-impact velocity $z^{1+} = \dot{y}^{+}\rho A\omega_1 l/F_{in}$ are presented. As we can see, the velocity can also increase after the impact $z^{1+}(t)$. This is connected with the accumulation of energy of the VIM elastic links.

Vibro-impact vibrations in the presence of harmonic excitation are defined by impact frequency characteristics of the amplitude. The dependence of maximum amplitudes of post-impact rebound of the microcantilever is presented in Fig. 3.28. The curves were obtained for the case when the excitation force was applied at different points of the microcantilever x_0/l, where x_0 – distance from the site at which the excitation force was applied to the link fixing site. Parts of curves in dash lines correspond to the areas of unsettled laws. The shown curves have maximum values in the case of resonance when the excitation frequency is close to the natural frequencies of the structure.

In the cases when the site at which the force was applied is close to the essential points of the second or the third mode of vibration, in the presence of high

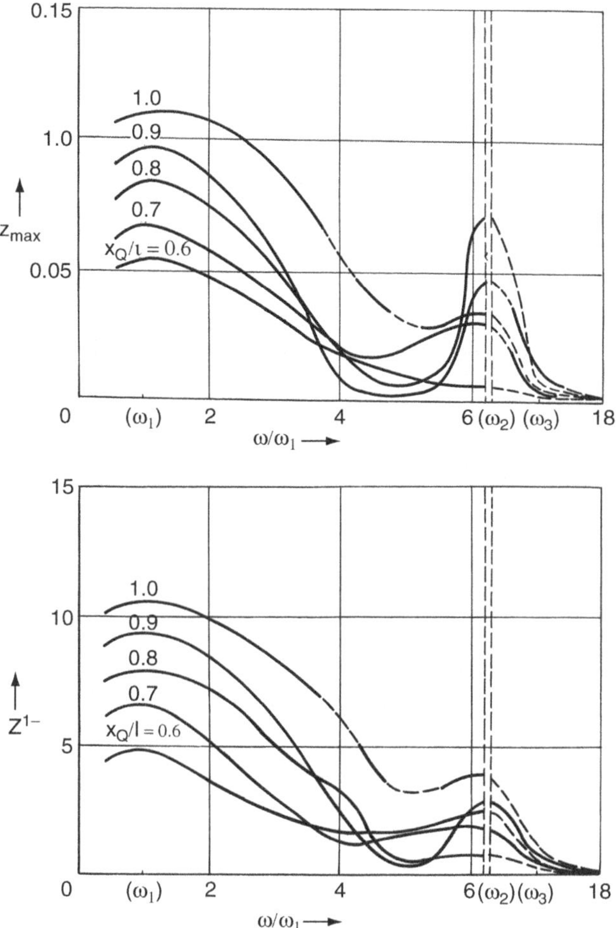

Fig. 3.28 Dependencies of amplitude (*top*) and impact velocity (*bottom*) frequency characteristics upon the excitation force application site

excitation frequencies, the link rebound amplitudes may be changed and the areas of settled vibration laws may be expanded. This is explained by excitation diagrams the amplitudes of the second or third mode of vibration are increased and at the same time dissipation of energy in the material of the structure increases, therefore the post-impact rebound amplitudes decrease. The amplitudes of the structure excited in this matter increase when the excitation frequencies are close to the second natural frequency. It is interesting to note that in the case when the site at which force was applied coincides with the essential points of the third mode of vibration or is close to them and when excitation frequencies are higher than the second natural frequency, settled periodic impact vibrations are obtained, which is represented by respective parts of curves in continuous lines. Thus, by applying the

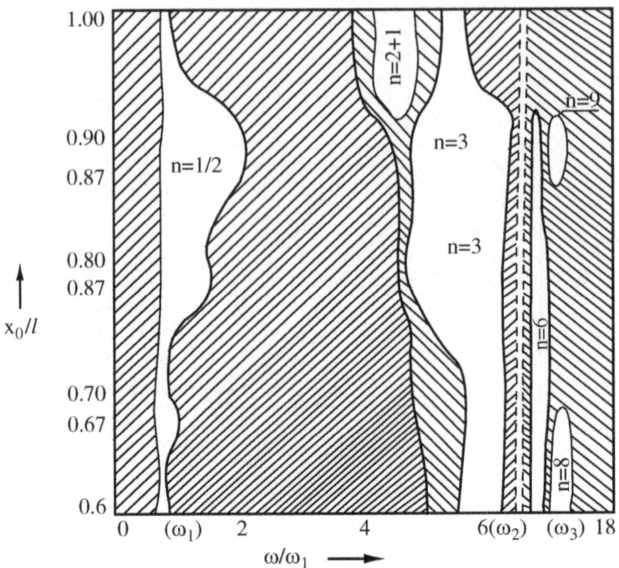

Fig. 3.29 Dependence or the existence of diverse law areas upon the position of the support

excitation force to the particular points of the third vibration mode of the elastic structure, the speed of VIM operation can be increased.

Figure 3.29 illustrates the possibility to expand the area of settled laws by changing the location of the support (x_0/l).The area of periodic vibration law (right-hand dashes) expands upon support displacement from the free end of the microcantilever towards its fixing point. Additionally, tendencies are seen towards the expansion of the areas of existence and the appearance of n-recurrent areas when the position of the support coincides with the particular points of the second or third mode of vibration ($x_0/l = 0.67, 0.78, 0.87$).Thus, the expansion of area $n = 1/2$ at the said points confirms that such law is related to an increase in the vibrations of the second or third mode of the elastic link.

Second support enhances the functional possibilities of the VIM. As an illustration of such VIM operation, Fig. 3.30 provides the phase trajectory of vibration of the free end of the elastic microcantilever between support 1 and support 2, at the excitation frequency $\omega = 2\omega_1$, and upon change of the point at which excitation force is applied x_0/l. As can be seen from the chart, irrespective of the VIM symmetry, the velocities before the impact on support 1 and support 2 are not equal $z^{1-} = \dot{y}^{-}\rho A\omega_1 l/F_{in}$ and $z^{1+} = \dot{y}^{+}\rho A\omega_1 l/F_{in}$. This can be explained by the fact that already at the initial moment of applying the excitation force, the elastic link starts to move from the equilibrium position and before the impact on the support passes a distance equal to the gap between the impact surfaces. Thus, the velocity of the first impact of the elastic link on the support is smaller than that of other impacts, and hence the energy dissipation resulting from such impact, being proportional to the velocity of impact, is also lower.

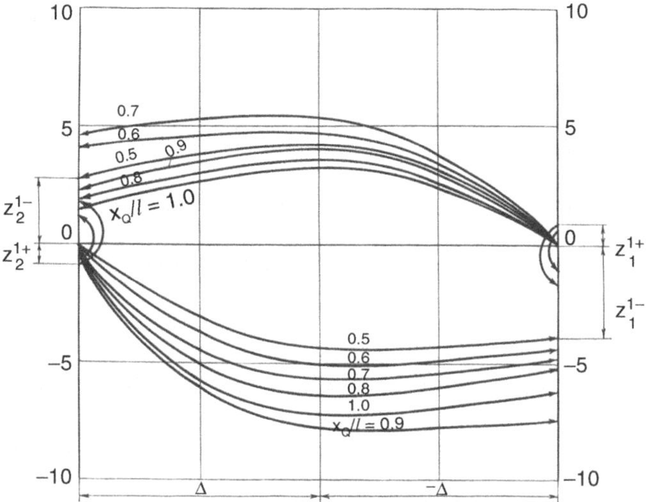

Fig. 3.30 Phase trajectory of the microcantilever motion between two supports

In addition to the sine excitation, other ways of harmonic excitation are used in VIM. Figure 3.31 provides areas of settled vibration laws in the cases of half-sine, saw shaped or square excitation that enable to select the best diagrams of VIM control. When these are compared to the areas presented in Fig. 3.26, it is evident that the most effective speed of operation is ensured by using sine excitation for the control of VIM.

One of the main requirements in VIM is its speed of operation. The speed of operation can be increased by minimizing the mass of the elastic structure and by selecting a respective control diagram. Besides, the speed of operation can be increased by selecting the configuration of the elastic link. In addition to others, it is necessary to distinguish optimal configurations that tend to self-excitation by prescribed frequencies due to the effect of the impact, as described above. Such configurations obtained for the first (I), second (II) or third (III) frequencies of the free transverse vibrations are presented on the left-hand side of Fig. 3.32. As can be seen, among the areas presented on the right-hand side of Fig. 3.32, in the presence of low sine excitation frequencies, the area of settled periodic law (right-hand dashes) is the widest for a structure that is optimal according to the first frequency (Fig. 3.49). This can be easily explained as the transverse vibrations of such VIM are of the first mode. When the excitation frequency is increased, the periodic vibro-impact law becomes almost settled (points and dashes), with its areas alternating with unsettled (left-hand dashes) recurrent vibration laws. The structures with configurations obtained for the II (Fig. 3.50) and III (Fig. 3.51) natural frequencies do not have wide areas of settled laws in the presence of small excitation frequencies. They become wider when excitation frequencies exceed the second natural frequency ω_2. As was expected, the microcantilever whose optimal configuration III was obtained for frequency ω_3 has a settled vibration impact law till the

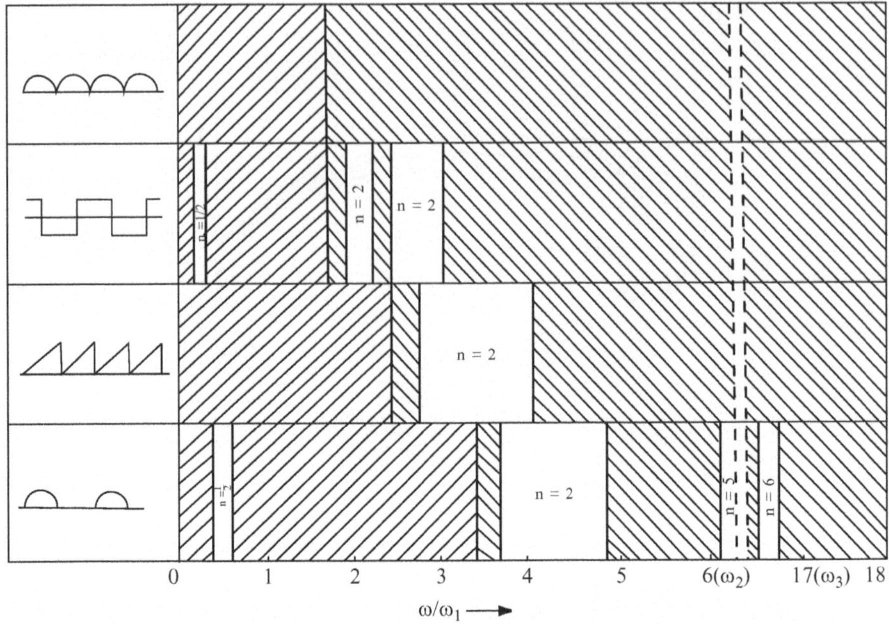

Fig. 3.31 Diverse vibro-impact motion law areas at different shapes of excitation pulse

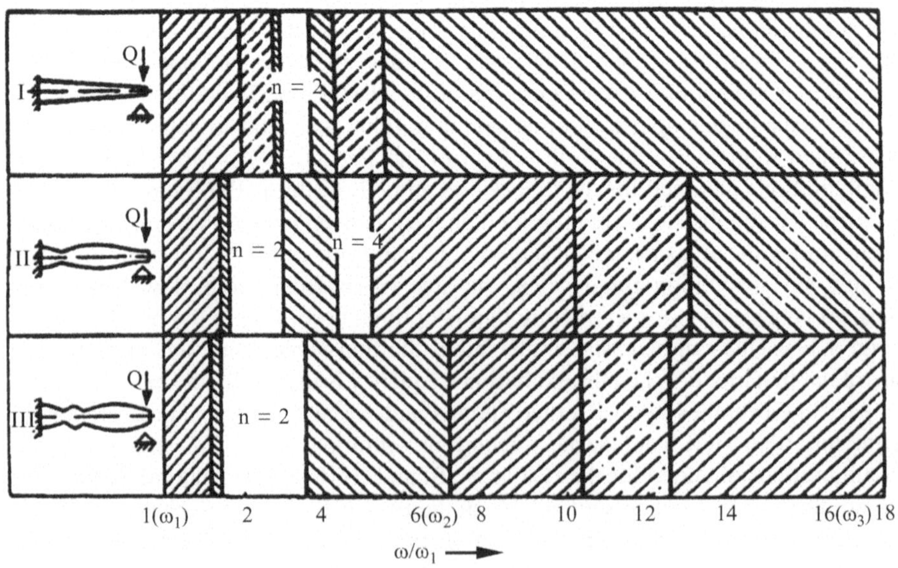

Fig. 3.32 Diverse vibro-impact motion laws areas of optimal microcantilevers

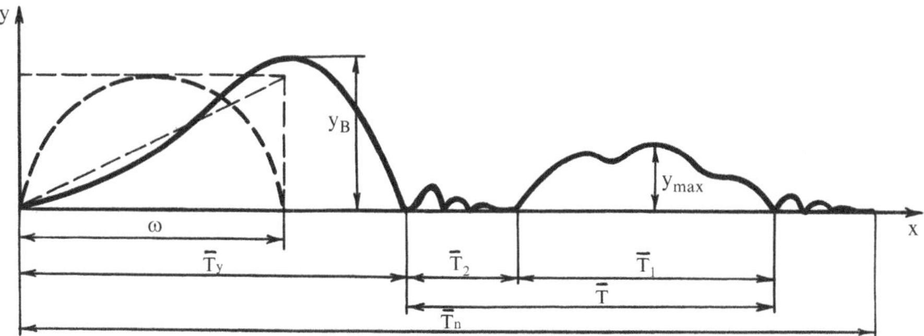

Fig. 3.33 Vibro-impact motion law of the microcantilevers at pulse excitation

excitation frequency reaches the third natural frequency. It is interesting to note that similar results also apply to structures obtained in the optimal design and with account to higher prescribed frequencies of transverse vibrations as well as structures with either longitudinal or torsional vibrations.

Most often the elastic structures of VIM are excited by discrete pulses of different shapes. The trajectory of vibro-impact motion is intricate and depends on the shape and amplitude of the excitation pulse (half-sine, square and triangular pulses are represented in dashed lines in Fig. 3.33) and on the pulse duration $T_i = 1/\omega$ and is defined by the excitation amplitude y_e, maximum post-impact rebound amplitude $y_{\max}$, duration of vibro-impact motion $\bar{T}_n$ the starting moment of the first impact $\bar{T}_s$, vibro-impact period $\bar{T}$, that consists of rebound duration $\bar{T}_1$ and impact duration $\bar{T}_2$. Besides, the impact-frequency and force characteristics are also important.

Dependencies of time characteristics of unstressed VIM excited by half-sine pulses on the position of the support x_0/l are presented in Fig. 3.34. The curves of vibro-impact motion duration $\tau_v = \omega_1 T_v$ are defined by minimal values obtained when the supports are located in the particular points of the third mode of vibration ($x_0/l = 0.67; 0.87$) and are identical to respective characteristics of VIM obtained in the presence of free vibro-impacts (Fig. 3.4). This allows concluding that basing on the free vibro-impact motion law of the VIM with elastic links it is possible to judge about other types of movement of such system. Irrespective of the support position, the time till the first impact $\bar{\tau}_s = \omega_1 \bar{T}_s$ only depends on the pulse duration and decreases upon its decrease (dash lines). The remaining time characteristics$\bar{\tau} = \omega_1 \bar{T}$, $\bar{\tau}_1 = \omega_1 \bar{T}_1$ and $\bar{\tau}_2 = \omega_1 \bar{T}_2$ are not that sensitive to the changes in the support position though some of them can be increased, too.

The position of the support has an effect not only on the time characteristics of VIM. In Fig. 3.35 the dependencies of excitation amplitude $z_e = y_e/l$ (continuous lines) on half-sine pulse duration (ω/ω_1), for different positions of the support show that upon shortening of the excitation pulse , the excitation amplitude markedly decreases. This is logical because short pulses of excitation are not sufficient to overcome inertia forces of elastic links, therefore they are slightly deflected.

The dependencies of contact pressure forces $P = F_y/F_{in}$ on the excitation pulse duration represented by dash lines show that when the support is located at the free

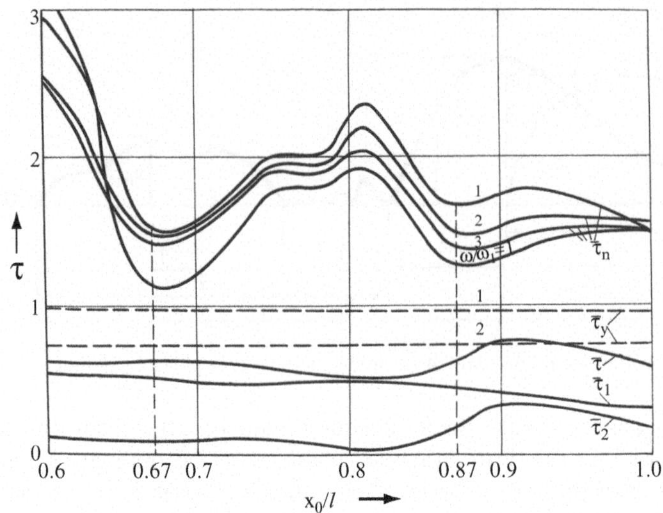

Fig. 3.34 Time characteristics of the VIM at pulse excitation

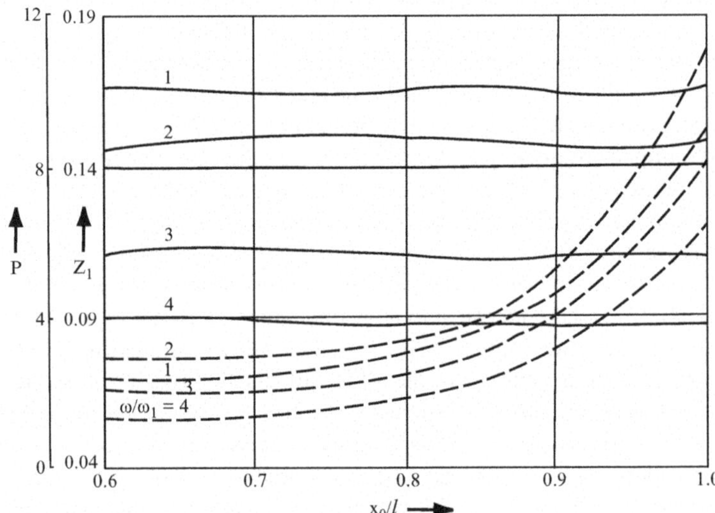

Fig. 3.35 Dependence of excitation amplitude and contact pressure force upon the position of the support at pulse excitation

end of the microcantilever, the contact pressure force, in discrete pulse excitation, is maximum.

When the configuration of the microcantilever is changed in the case of pulse excitation, the maximum amplitudes of post-impact rebound can be changed. The curves in Fig. 3.36 illustrate the dependence of rebound amplitudes of optimal

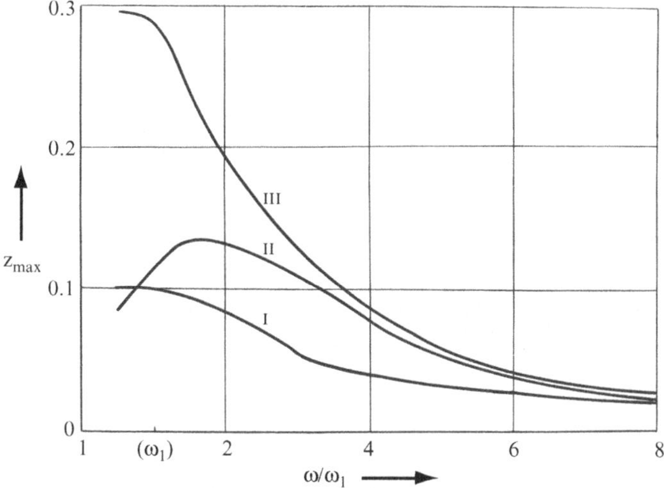

Fig. 3.36 Dependence of optimal microcantilever rebound amplitudes upon the duration of excitation pulse

configuration links obtained for the I, II and III frequencies $z_{max} = y_{max}/l$ on the excitation pulse duration.

3.7 Vibrational Stability of Vibro-Impact Microsystems

Vibro-impact microsystems usually operate in equipment subjected to other external mechanical effects, such as vibration, impacts, etc. Quite often external mechanical effects disturb the normal functioning of the elastic links of a VIM. If VIM functions in moving or acoustic equipment, the characteristics of vibrations operating in such equipment alter in certain intervals. For instance, wheeled vehicles excite vibrations of (2–5) and (20–60) Hz, caterpillar vehicles – (400–700) Hz, and ships – (2–35) Hz. The vibrations of flying apparatus are of (3–500) Hz frequency, up to 3.8 mm amplitude, (0.2–20) g acceleration. Acoustic vibrations up to 130 kHz appear when the velocity threshold level is exceeded.

A VIM affected by kinematic excitation is described by the same equation of dynamics

$$[M]\{\ddot{y}(t)\} + [C]\{\dot{y}(t)\} + [K]\{y(t)\} = \{Q(t)\}, \tag{3.36}$$

where $\{Q(t)\}$ – vector of external forces, $\{y(t)\}$ – vector of generalized displacements in the presence of kinematic excitation that is divided into two vectors of the following form:

$$\{y(t)\} = \begin{Bmatrix} \{y_S\} \\ \{y_K\} \end{Bmatrix}, \quad (3.37)$$

where $\{y_S\}$ – unknown vector of displacement, $\{y_K\}$ – prescribed vector of kinematic excitation.

By introducing the vector of state variables (3.37) into the Eq. (3.36), we obtain:

$$\begin{bmatrix} [M_{SS}] & [M_{SK}] \\ [M_{KS}] & [M_{KK}] \end{bmatrix} \begin{Bmatrix} \{\ddot{y}_S\} \\ \{\ddot{y}_K\} \end{Bmatrix} + \begin{bmatrix} [C_{SS}] & [C_{SK}] \\ [C_{KS}] & [C_{KK}] \end{bmatrix} \begin{Bmatrix} \{\dot{y}_S\} \\ \{\dot{y}_K\} \end{Bmatrix} + \begin{bmatrix} [K_{SS}] & [K_{SK}] \\ [K_{KS}] & [K_{KK}] \end{bmatrix} \begin{Bmatrix} \{y_S\} \\ \{y_K\} \end{Bmatrix} = \begin{Bmatrix} \{Q_S\} \\ \{Q_K\} \end{Bmatrix}, \quad (3.38)$$

where $\{Q_S\}$ – prescribed effect of the force, $\{Q_K\}$ – reacting force.

The first line of the equation of matrices (3.38) can be rewritten as:

$$\begin{aligned} &[M_{SS}]\{\ddot{y}_S(t)\} + [C_{SS}]\{\dot{y}_S(t)\} + [K_{SS}]\{y_S(t)\} \\ &= \{Q_S(t)\} - [M_{SK}]\{\ddot{y}_K(t)\} - [C_{SK}]\{\dot{y}_K(t)\} - [K_{SK}]\{y_K(t)\}, \end{aligned} \quad (3.39)$$

when the initial conditions

$$\{y_S(0)\} = \{\dot{y}^0{}_S\}, \{y_S(0)\} = \{y^0{}_S\}. \quad (3.40)$$

By applying this algorithm, data was developed and calculations of the VIM vibratory stability were carried out. The position of the support that is in mutual contact with the microcantilever exerts a marked influence on the VIM vibratory stability.

This is confirmed by the dependences of limiting vibratory stability $k = y_0\omega^2/g$ on the kinematic sine excitation frequency ω.

As we can see in the chart, the most stable in respect of vibrations (i.e. maintaining normally connected links) in the presence of highest excitation frequencies ω/ω_1 is such a VIM structure in which the support is connected at the distance $x_0/l = 0.78$. This is the nodal point of the second transverse vibration mode, and, as shown above, such VIM with a support in this point has the highest value of the first natural frequency ω_1 (Fig. 3.37).

Thus, when the support is located at the nodal point of the second mode, all conditions are created for the dissipation of the kinematically transferred energy in the structure material, thereby retaining it on the support. Another point for the location of the support ensuring the maintenance of the VIM vibrational stability is the point of the maximum amplitude of the third vibration mode $x_0/l = 0.67$.

Most dangerous in VIM are such kinematic excitation frequencies that coincide with the first natural frequency of the structure ω_1. Figure 3.38 provides the dependences of the vibration amplitudes of the microcantilever contacting a semi-rigid support, when the kinematic excitation frequency is equal to the first natural

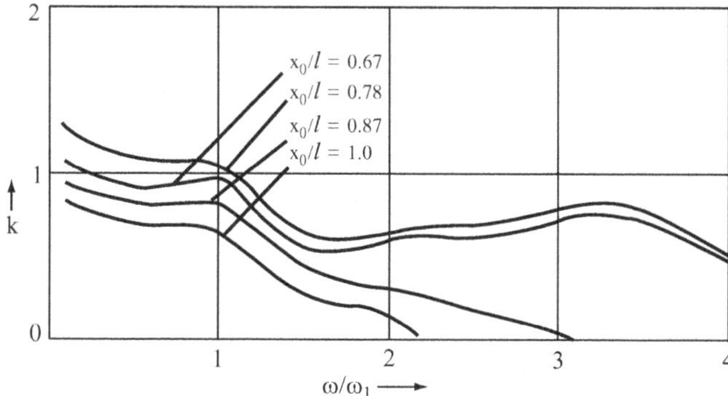

Fig. 3.37 Dependence of the VIM vibrational stability upon the position of the support in the case of kinematic excitation. The most stable in respect of vibrations (i.e. maintaining normally connected links) in the presence of highest excitation frequencies ω/ω_1 is such a VIM structure in which the support is connected at the distance $x_0/l = 0.78$

Fig. 3.38 Dependence of the microcantilever vibration amplitude upon the position of the support

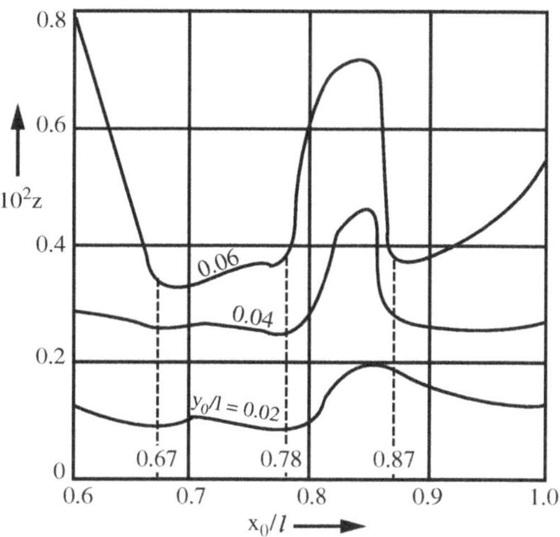

frequency of the structure. We can see from the diagrams that in the presence of small amplitudes of the excitation signal ($y_0/l = 0.02$) the microcantilever contacts the support. The amplitudes of the microcantilever decrease due to the dissipation of the kinematically transferred energy in the structure material when the support reaches the points $x_0/l = 0.67$ and 0.87. Upon increase of the amplitudes of kinematic excitation $y_0/l > 0.04$, the microcantilever starts to rebound off the support. This is confirmed by the dependences of the contact pressure force provided in Fig. 3.39. Upon increase of prestress, the VIM vibrational stability increases, however, additional energy is needed for the control of the VIM.

In practical usage VIMs are often subjected to impact loads. Due to the effect of the impact, in a short period of time the velocity and amount of motion change. In the

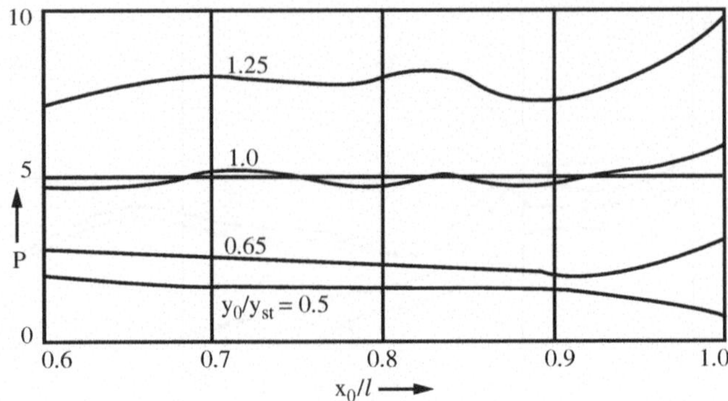

Fig. 3.39 Dependence of the contact pressure force upon the position of the support

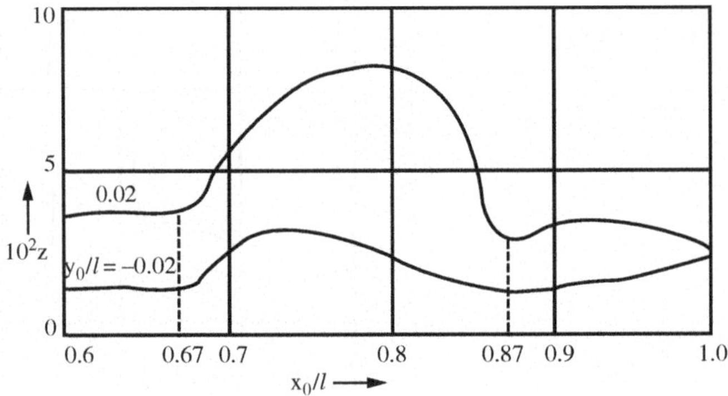

Fig. 3.40 Dependence of the microcantilever rebound amplitudes upon the impact excitation direction

modeling it is important to determine the influence of the simulated impact direction on the rebound amplitudes of the VIM links. Figure 3.40 provides the dependences of the microcantilever rebound amplitudes upon the direction of the impact impulse $y_0/l > 0$ when the direction of its effect pushes the elastic link off the support, and $y_0/l < 0$ – when it is the opposite case. The presented curves show that the rebound amplitudes, when $y_0/l > 0$, are markedly higher that the rebound amplitudes of the links effected by the opposite direction of the impact. Besides, the amplitudes decrease when the support is located in the particular points of the third vibration mode.

The curves in Fig. 3.41 demonstrate the influence of changes in the variables of the VIM structure: the length of the microcantilever l, the support position x_0 and applied mass $\bar{m}$. Figure 3.42 illustrates the effect of the microcantilever bending at angle α on its impact stability.

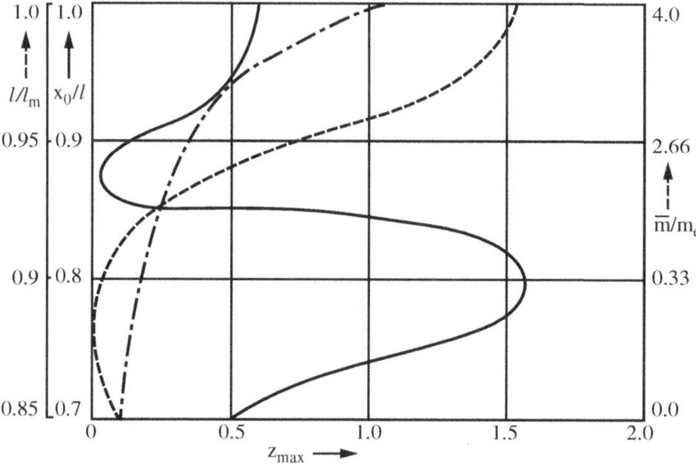

Fig. 3.41 Dependence of the microcantilever rebound amplitudes upon the VIM parameters

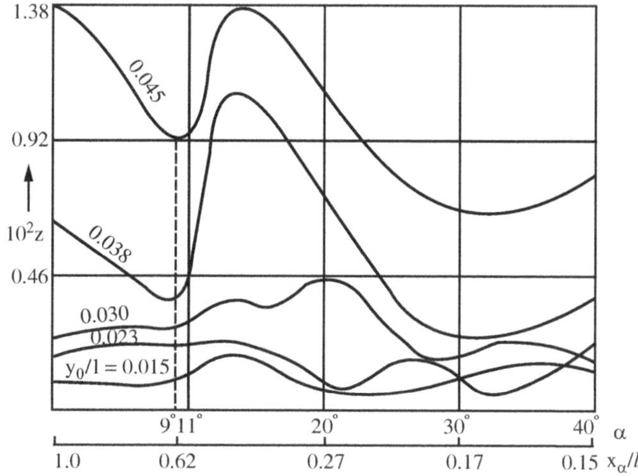

Fig. 3.42 Dependence of the microcantilever rebound amplitudes upon the bending angle

3.8 Diagnostics of Parameters of Vibro-Impact Microsystems

Vibro-impact microsystems used in complex systems have to meet high quality requirements. The issues of ensuring long-term functioning alongside with high reliability and specifying the service life under certain conditions of operation have not been adequately studied. The main requirements facing a designer are to create such VIMs that would be capable of operating for a few decades and to develop methods allowing to specify longevity and service life.

The analysis has shown that most often VIMs fail when defects appear in the elastic links. Such defects may be of different nature but most often they are related to a non-uniform distribution of the structure material density, elasticity modulus and deviations of geometrical parameters. According to the degree of localization, defects of finite dimensions and point-shaped defects may be distinguished. The reasons as to why VIMs get out of order before the specified time are most often related to structure damages in the course of production or exploitation, overloading and fatigue. The structure disintegrates more rapidly under the effect of vibrations.

The developed VIM models enable to simulate the structure defects decrease or increase of the cross-section. The studies enable to determine the effect of such defects on the VIM dynamics and, additionally, to use the structures with changed cross-section for the improvement of the VIM dynamic properties.

The amplitude characteristics of structures with changed cross-sections are presented in Fig. 3.43. The curves in Fig. 3.43 (top), illustrate the dependences of the ratio between maximum rebound amplitudes of the microcantilever with finite elements of a twofold decreased cross-section $y_{\bar{k}}$ on the one hand, and rebound amplitudes of the initial structure of constant cross-section, y_{in} on the other hand, and dependences of the ratio between the first three (I, II, III) natural frequencies of the microcantilever of decreased cross-section ω_{ik} on the one hand, and the respective frequencies of the initial structure of constant cross-section ω_{in} on the other hand. The structure was meshed by $m = 10$ finite elements. The values for the finite element $m = 11$ coincide with the values of the initial structure of constant cross-section. We can see from the curves that the amplitudes of free impact vibration of the microcantilever may be markedly decreased by decreasing the cross-sections of the finite elements $m = 3$, $m = 5$ and $m = 9$ and they may be increased by decreasing the cross-section of the finite element $m = 10$ in whose nodal section the support is located. The curve of amplitude dependences is similar to the curve of the II natural frequency.

The natural frequencies of transverse vibrations are lower than the frequencies of the initial structure and exceed them only provided the cross-section of the free end of the microcantilever is decreased. Figure 3.43 (bottom) presents curves obtained upon a twofold increase of separate cross-sections of the microcantilever and their displacement along the axis. We can see that the rebound amplitudes of the microcantilever, $y_{\bar{m}}$ can be decreased by increasing the cross-sections of the finite elements $m = 3$, $m = 5$ or $m = 10$. The amplitudes increase when the finite element $m = 6$ is made thicker. The natural frequencies of such structure are higher than the frequencies of the initial structure when the increase of the cross-section of the microcantilever occurs nearer to the fixing site, and lower when the free end is thicker.

We will analyze the reasons of the above. The decrease in the rebound amplitude, occurring upon decrease of the cross-section of the finite element $m = 3$ is attributed to the fact that it is in the finite element $m = 3$ ($x/l = 0.24$) that the optimal configuration of the microcantilever has minimum cross-section for the given second frequency of natural transverse vibrations (Fig. 3.50). Hence the decrease in the rebound amplitudes can be explained by intensification of the amplitudes of the second mode of vibrations, whereupon dissipation of energy in the structure material increases.

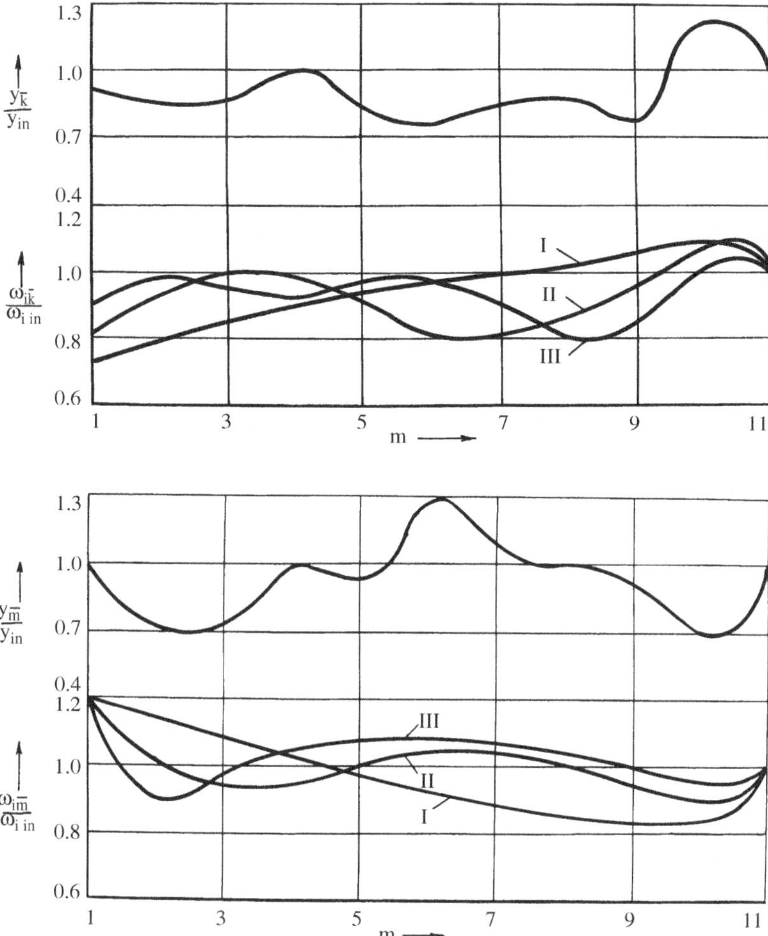

Fig. 3.43 Dependences of rebound amplitudes of microcantilevers of reduced (*top*) and increased (*bottom*) cross-sections

To explain the reasons of the decrease in the rebound amplitudes upon change of other cross-sections, the simulated microcantilever must be analyzed as a structure consisting of two parts separated by the changed cross-section.

The computational model for the microcantilever with a defect separating it into two parts 1 and 2 connected in point B is presented in Fig. 3.44. Displacements are denoted as y_1, and y_2 and the frequencies of the parts 1 and 2 are respectively ω_1 and ω_2. Vibro-impact motion of the part 2 evoke parametric vibration of the part 1. If an artificial pin-connection is installed at the site of the changed cross-section, the part 1 will be equivalent to a pin-supported microcantilever, and the part 2 to a microcantilever. Hence the natural frequencies ω_1 and ω_2 of the parts are respectively determined as for a pin-supported and non-supported microcantilever.

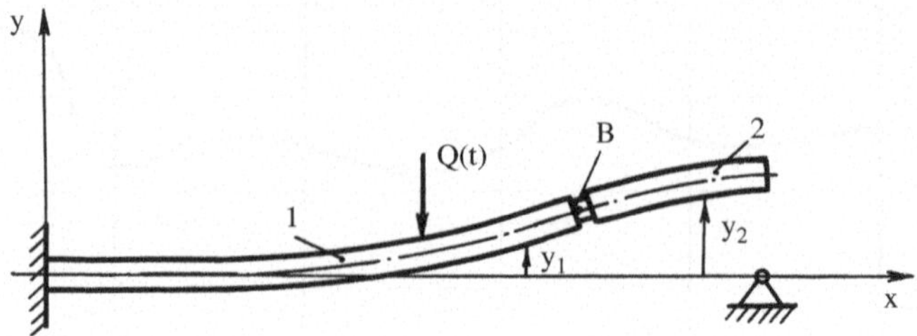

Fig. 3.44 Computational model of the microcantilever of modified cross-section

An increase or decrease of the rebound amplitudes of the structure in the process of vibro-impact oscillations is related to parametric vibrations. In the cases when the frequencies of transverse vibrations of the parts 1 and 2 are equal, $\omega_1 = \omega_2$ we have a case of internal resonance and, as a result, the structure rebound amplitudes should increase. This happens when the cross-section of the finite element $m = 6$ increases or that of the finite element $m = 7$ decreases. In the cases when $\omega_1 = 2\omega_2$ or $2\omega_1 = \omega_2$ we obtain the most favorable conditions for a high energy dissipation in the structure material because the vibrations of its parts are pre-phasic, which increases the amplitudes of higher vibration modes. The case $\omega_1 = 2\omega_2$ corresponds to a decrease in the cross-section of the finite element $m = 6$ or increase in the cross-section of the finite element $m = 5$ and the case $2\omega_1 = \omega_2$ corresponds to a decrease in the cross-section of the finite element $m = 9$ or an increase in the cross-section of the finite element $m = 10$. The increase in rebound amplitudes, upon a decrease of the cross-section of the finite element $m = 10$ is related to the decreased stiffness of the contacting part.

Thus, the phenomena of the appearance of parametric vibrations may be used to explain the causes of the change in the VIM dynamic characteristics and to make an accurate diagnosis of the structure state.

The characteristics of impact frequency amplitude obtained in the case of harmonic sine excitation of the microcantilever with a smaller cross-section are represented in Fig. 3.45 by continuous lines and in the case of half-sine excitation by dashed lines. Depending on ω/ω_1 and the number of the finite element of smaller cross-section, the decrease in the post-impact rebound amplitudes recurs several times (Fig. 3.45 (top)). It is interesting that the curves obtained for non-resonance vibrations ($\omega/\omega_1 = 1$), by their nature coincide with the curve of free impact vibrations. Maximum rebound amplitudes are characteristic of a structure excited by resonance frequencies $\omega/\omega_1 = 1$. In all cases of harmonic excitation, when the excitation frequency exceeds that of the resonance frequency $\omega/\omega_1 > 1$, upon decrease of the cross-section of the finite element $m = 3$ minimums appear both in the amplitude frequency characteristics (Fig. 3.45 (top)) and in the impact frequency characteristics (Fig. 3.45 (bottom)). It is interesting to note that in the case of excitation by half-sine pulses a wider range of amplitude changes appears, and this is another confirmation that the reasons of such change are related to the ratio between the transverse vibration frequencies of the parts of the structure.

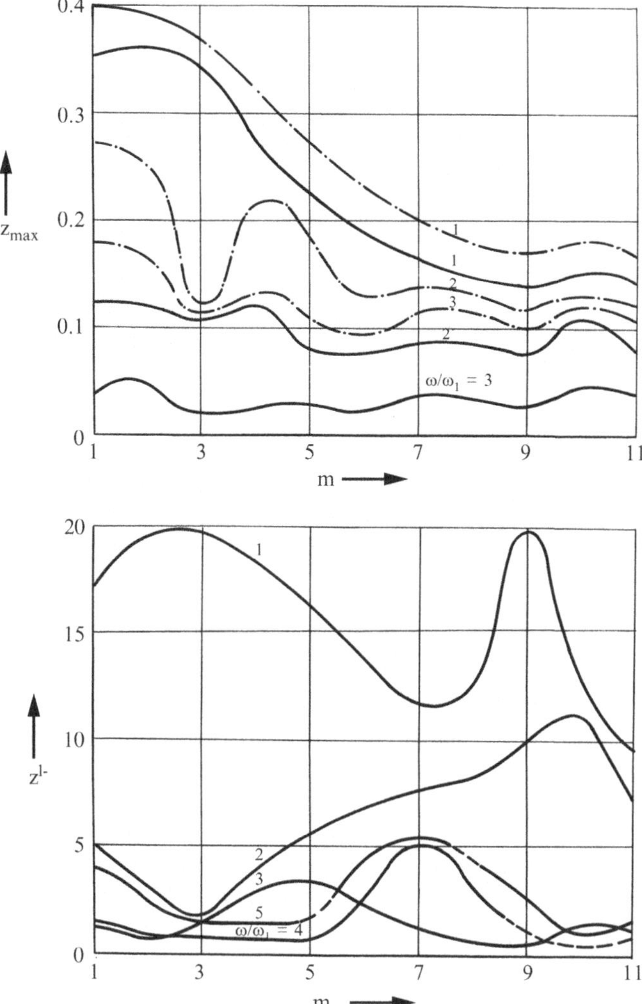

Fig. 3.45 Amplitude (*top*) and impact velocity (*bottom*) frequency characteristics of the micro-cantilever of reduced cross-section

By increasing the excitation frequency, unsettled regimes of vibro-impact motion appear (dashed parts of the curves in Fig. 3.45 (bottom)).

Figure 3.46 provides n-recurrent settled motion laws of an elastic structure of decreased cross-sections in the presence of excitation by a sine pulse. As the number of the finite element of decreased cross-section, m increases, the area of settled periodic motion law, in which during one excitation period the link makes one impact on the support, expands. By increasing the excitation frequency, we see an increase in the n-recurrent motion law number, which, for example, $n = 4 + 3$, means that in the course of four excitation periods the structure makes one impact

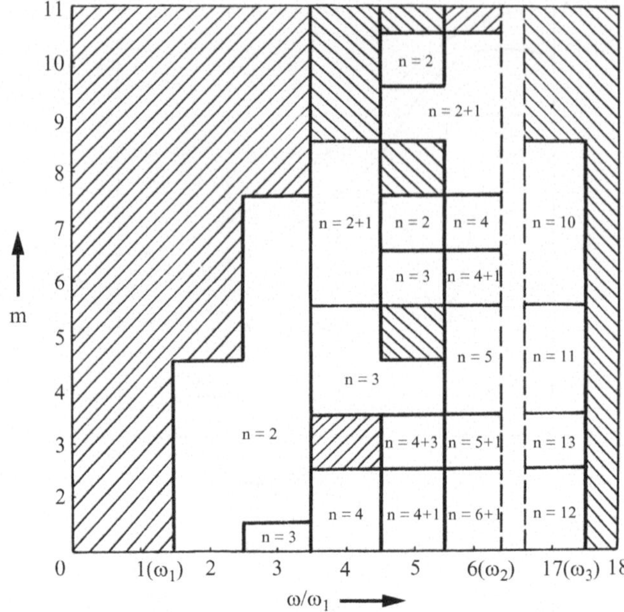

Fig. 3.46 *n*-recurrently -settled motion laws of the microcantilever of smaller cross-sections

against the support, and then three successive impacts occur in the course of three excitation periods. Thus, to ensure the desirable law of the VIM, it is enough to decrease the cross-section of the structure at a certain place. The left-hand dashes represent the areas of unsettled motion law.

By exciting the structure of changed cross-sections by discrete pulses, it is possible to decrease or increase the rebound amplitudes. Figure 3.47 provides the amplitude frequency characteristics of the structure of decreased cross-sections excited by discrete half-sine shaped pulses. The provided curves are defined by a decrease in the rebound amplitudes when the cross-sections of finite elements $m = 3$; $m = 6$ or $m = 10$ decrease. Rebound amplitudes increase upon decrease of the cross-sections of the finite elements $m = 5$ or $m = 8$.

In most cases of VIM design, the possibility of the VIM functioning and viability under the effect of extra big VIM overloads has to be predicted. Such overloads appear when explosions or accidents effect the equipment in which the VIM operates. It is not possible to check the influence of such effect experimentally, as there is no appropriate equipment, whereas by using mathematical models overloads can be easily simulated. Figure 3.48 provides the dependences of the maximum amplitudes of the structure under the kinematic effect of overload $k = y_0\omega^2/g$. The dashed line $z_{max} = 0.22$ represents the medium structure part deflection that evokes tension corresponding to the yield limit of the structure material. When the values of $z_{max} = 0.22$ are exceeded, the tension of the structure material is considered to be residual. Thus, we can see from the diagram that in the presence of resonance excitation $\omega/\omega_1 = 1$, the VIM fails when $k = 11 \times 10^3$. In the cases

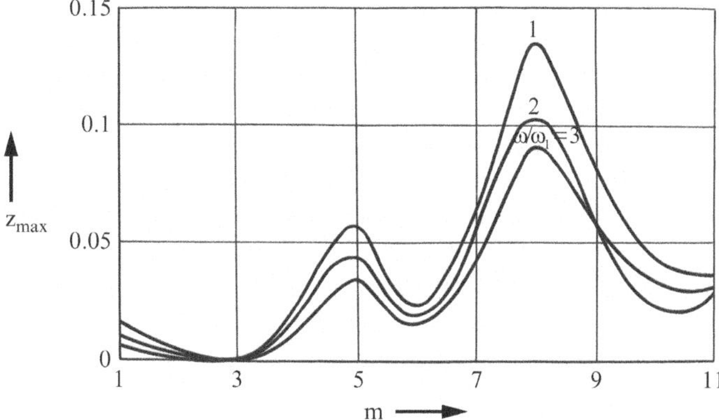

Fig. 3.47 Frequency characteristics of the amplitude of the microcantilever of reduced cross-sections at discrete pulse excitation

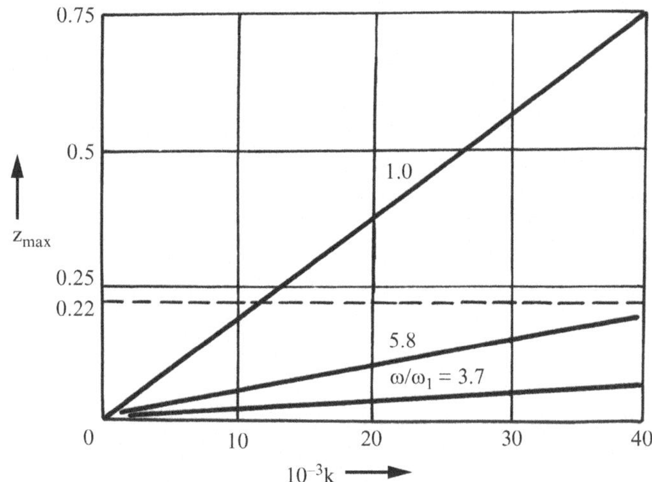

Fig. 3.48 Dependence of the vibration amplitudes of the microcantilever medium part upon the overload accelerations

when the frequency of kinematic excitation exceeds the resonance frequency, i.e. when $\omega/\omega_1 = 3.7$ and $\omega/\omega_1 = 5.8$, the VIM endures higher overloads.

3.9 Optimization Results for Structures Undergoing Periodic and Transient Vibrations

The aim of optimization is to select such geometrical parameters that would correspond to the technical characteristics of the system and give a minimum value to a certain quality functional. The structure optimized with reference to the established criteria and constraints may turn out to be irrational. For instance, if the technological requirements are neglected and not included in the constraints, there is a possibility that the developed structure will be optimal in terms of mass, but irrational in the technological aspect. But using volumetric etching of silicon technologies it is possible to obtain optimal configurations.

In the design of VIM, both numerous constraints and the purpose of the structure should be taken into account. As concerns the constraints, geometrical and structure performance constraints should first of all be distinguished. The target functions, most frequently used in optimal design, express the system mass minimized with account to the constraints of a prescribed vibration frequency.

3.9.1 Optimization Results of Structures Undergoing Periodic Vibrations

It is desirable to design such a structure, the natural frequencies of which would not be fitted in a certain interval ($\omega_{\min}$, $\omega_{\max}$).There are many reasons as to why the natural frequencies should not be fitted in an undesirable range, for example, in cases when it is known that the excitation frequency of the object in which the VIM will operate fits into the same interval, which may result in the appearance of resonance.

We will minimize the structure mass at a prescribed frequency of transverse vibrations. The target function of structure optimization is expressed as:

$$\Phi(A) = \min \rho a(A_1 + A_2 + \ldots + A_m),$$

where A_i – cross-sections of structure components, $i = 1, 2, \ldots, m$

When the method of nonlinear programming, e.g. gradient projection, is used, the inequality-shaped constraints are employed:

$$f_m(A) = 1 - A_m/A_i \leqslant 0,$$
$$f_{m+1}(\omega) = 1 - \omega/\omega^* \leqslant 0,$$

where ω^* – the prescribed frequency of structure vibrations.

By using the gradient projection method, optimal configurations of the micro-cantilever were obtained. Figures 3.49–3.51 provide optimal structures obtained for the prescribed frequencies of natural transverse vibrations ω_1, ω_2 or ω_3.

Figure 3.52 presents optimal structures obtained for the transverse vibration frequencies ω_4 and ω_5. As may be seen from the sketches, with the increase of frequency value, the number of minimum cross-sections increases. From sketches in Figs. 3.49 and 3.52, it is not difficult to measure the distances from the minimum and maximum cross-sections to the structure fixing site.

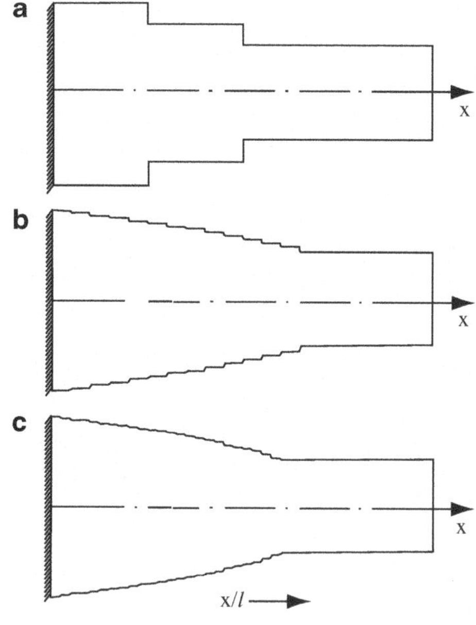

Fig. 3.49 Optimal microcantilevers at the first natural frequency of transverse vibrations. *Top*: $m = 4$. *Middle*: $m = 20$. *Bottom*: $m = 40$

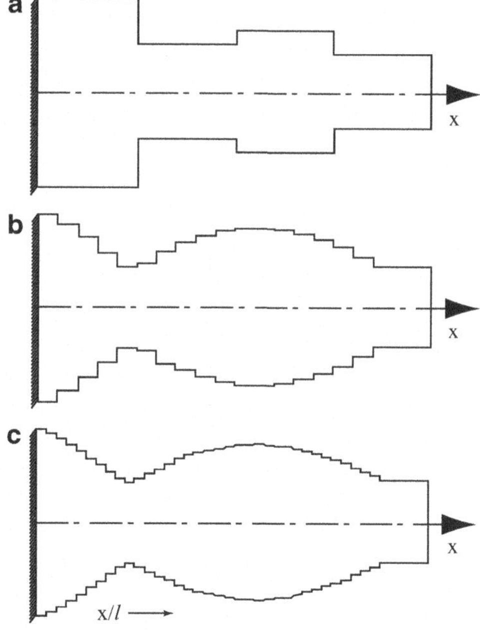

Fig. 3.50 Optimal microcantilevers at the second natural frequency of transverse vibrations. *Top*: $m = 4$. Middle: $m = 20$. *Bottom*: $m = 40$

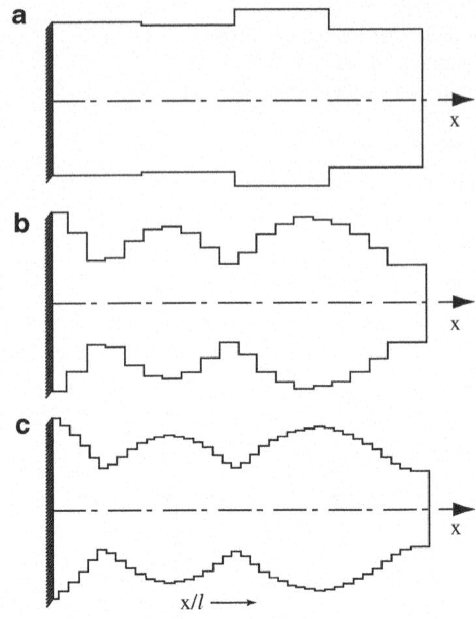

Fig. 3.51 Optimal microcantilevers at the third natural frequency of transverse vibrations. *Top*: $m = 4$. *Middle*: $m = 20$. *Bottom*: $m = 40$

For example, the minimum cross-section of the structure that is optimal at the second value of transverse vibration frequency ω_2 is at the distance of $0.24l$ from the fixing site, and at the third value – $0.15l$ and $0.5l$, respectively (Figs. 3.50 and 3.51), where l is the structure length.

The analysis of optimal structures presented in Figs. 3.49–3.52 shows that the recurrence of maximum and minimum cross-sections corresponds to the positions of particular (maximum amplitude and nodal) points in the modes of cross-section torsional vibration Φ_{ii} indicated in Fig. 3.2 (bottom). As can be seen from the sketches, minimum cross-sections of optimal structures correspond to the maximum amplitude points of the cross-section torsional vibration modes, and the maximum ones – to the nodal points of these modes.

In the VIM design practice, it is often necessary have not just one but several natural frequencies beyond a certain frequency range because when the structure is optimized in respect of one frequency, the other frequencies change considerably. For instance, for the structure optimal in respect of the second frequency ω_2 the first frequency ω_1 decreases by 21%, and for the structure that is optimal in respect of the third frequency ω_3, the second frequency ω_2 decreases by 18%, and the first frequency by 25%. This may have a negative effect on VIM, because when some of the natural frequencies are moved from the dangerous zones, the other may get to these zones. Figure 3.53 presents optimal configurations for the first ω_1 and the second ω_2 (Fig. 3.53 (top)), the first ω_1 and the third ω_3 (Fig. 3.53 (middle)), the second ω_2 and the third ω_3 (Fig. 3.53c) frequencies. The optimal structure for the first three natural frequencies is presented in Fig. 3.54.

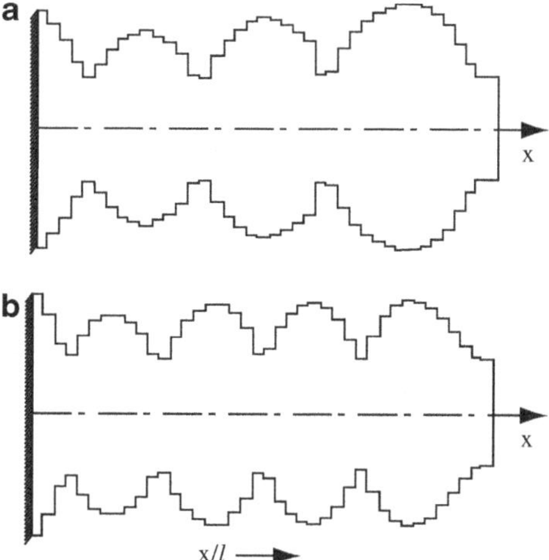

Fig. 3.52 Optimal microcantilevers at the fourth (**a**) and fifth (**b**) natural frequency of transverse vibrations

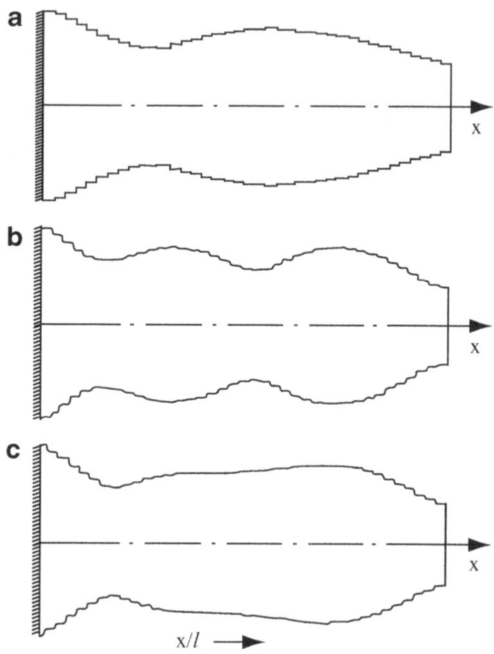

Fig. 3.53 Optimal microcantilevers at the frequencies. *Top*: $\omega_1\omega_2$. *Middle*: $\omega_1\omega_3$. *Bottom*: $\omega_2\omega_3$

In the VIM design practice, structures with minimum difference between the first and the second frequencies in the links are important. This enables to enhance the speed of operation and vibrational stability of VIM because the first natural frequency of such structure usually increases. Figure 3.55 presents an optimal

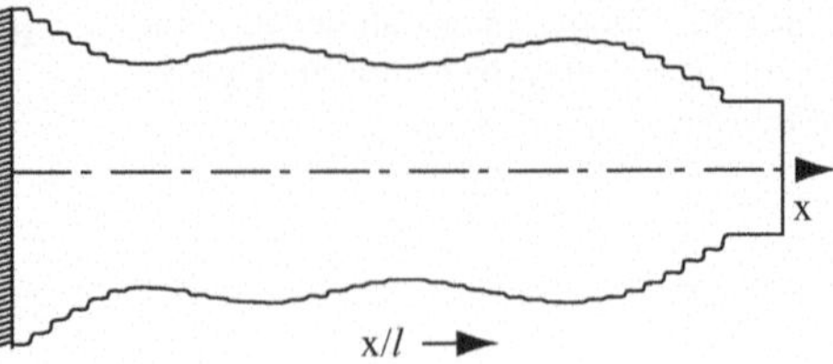

Fig. 3.54 Optimal microcantilever at ω_1, ω_2, ω_3 frequencies

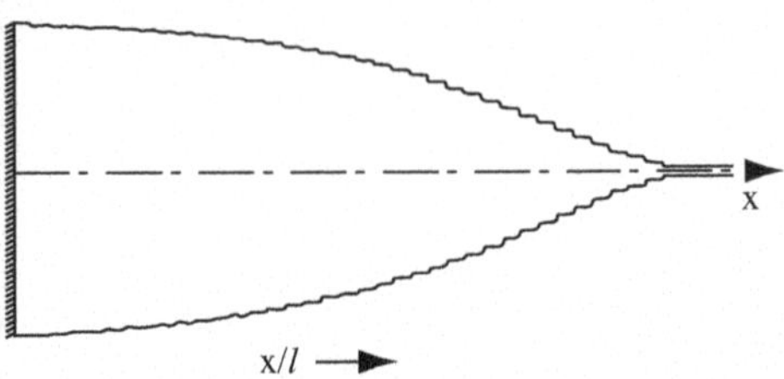

Fig. 3.55 Optimal microcantilever in which $\omega_2 = 1.23\omega_1$,

structure obtained upon constraining the first and the second natural frequency so that the difference between the values would be smallest possible. In the obtained structure the natural frequency ω_1 differs from the second natural frequency ω_2 by 23%, whereas in the initial structure of uniform cross-section the difference between such frequencies constitutes a 628%.

In cases when VIM are composed of structural elements fixed at both ends, the medium parts of such elements vibrate at a maximum amplitude. Optimal configurations of such structures for the first three natural frequencies ω_1, ω_2, ω_3 are presented in Fig. 3.56. Figure 3.57 provides optimal configurations obtained for the fourth ω_4 and fifth ω_5 frequencies of transverse vibrations.

In the design of optimal structures fixed at both ends, the prescribed frequencies were equal to respective frequencies of the initial structure of uniform cross-section. Such configurations are symmetrical in respect of the medium part of the structure. After optimization, at the first frequency ω_1 the structure mass, as compared with the initial mass, decreased by 23%, at the second frequency ω_1 by 13%, at the third frequency ω_3 by 12%, and at the fourth frequency ω_4 only by 11%.

VIM consists of a microcantilever contacting a support. In such structures, it is not only the optimal configuration but also the position of the support that matters. Optimal structures were designed in respect of the first three natural frequencies, for different location of supports (Figs. 3.58–3.60). In the optimization in respect of the first frequency (Figs. 3.58–3.59), the structure mass decreases by 16% when the support is located at the end of the microcantilever $x_0/l = 1$, where x_0 is the support distance to the fixed end (Fig. 3.58 (top)), by 17% when $x_0/l = 0.9$ (Fig. 3.58 (middle)), by 20% when $x_0/l = 0.8$ (Fig. 3.57 (bottom)), by 26% when (Fig. 3.59 (top)), by 31% when $x_0/l = 0.6$ (Fig. 3.59 (middle)) and by 33% when $x_0/l = 0.5$ (Fig. 3.59 (bottom)). The obtained optimal configurations have a decreased

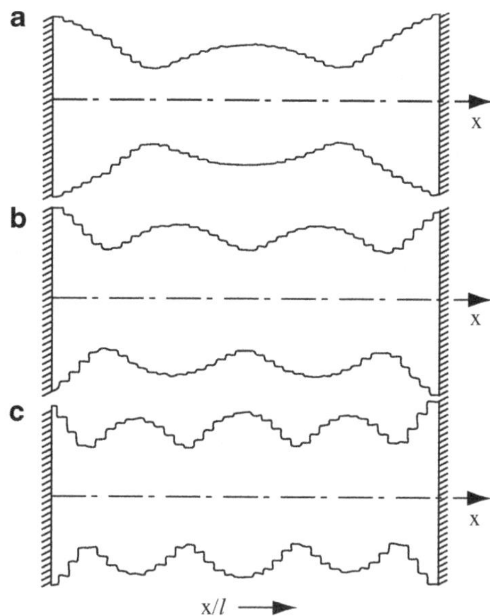

Fig. 3.56 Optimal structures fixed at both ends at different frequencies. *Top*: ω_1. *Middle*: ω_2. *Bottom*: ω_3

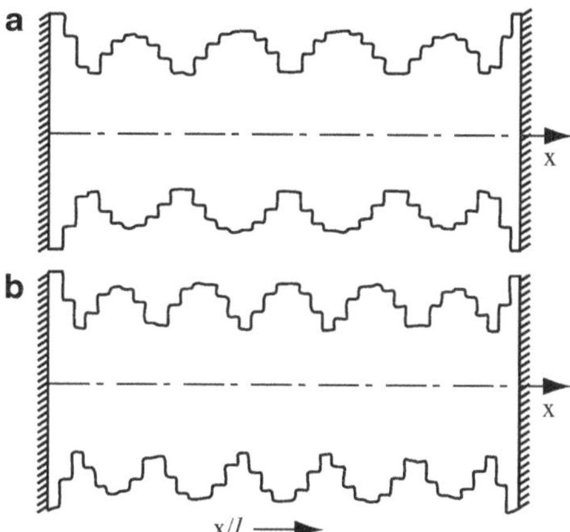

Fig. 3.57 Optimal structures fixed at both ends at different frequencies. *Top*: ω_4. *Bottom*: ω_5

cross-section at the distance $x_0/l = 0.15$ when the support is positioned in point $x_0/l = 0.5$, and at the distance $x_0/l = 0.32$, when the support is positioned at the free end of the microcantilever $x_0/l = 1$.

Optimal structures provided in Fig. 3.60 were obtained for the second prescribed frequency of transverse vibrations, when the support is at the free end of the microcantilever $x_0/l = 1.0$ (Fig. 3.60 (top)) at the distance $x_0/l = 0.9$ and at the distance $x_0/l = 0.8$ (Fig. 3.60 (bottom)). Each of the given optimal structures has two decreased

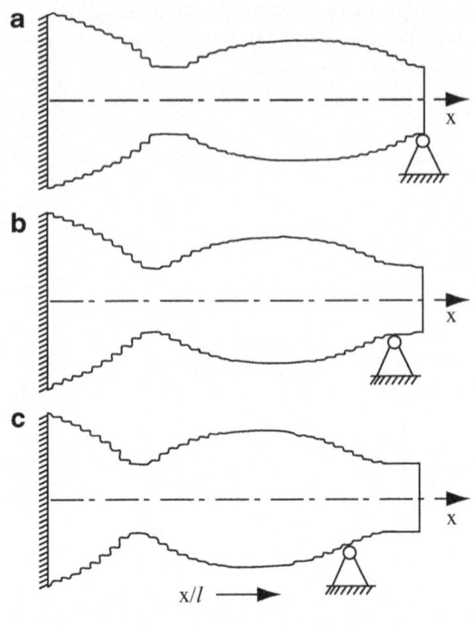

Fig. 3.58 Optimal supported microcantilevers at prescribed first natural frequency. *Top*: when $x_0/l = 1.0$. *Middle*: when $x_0/l = 0.9$. *Bottom*: when $x_0/l=0.8$

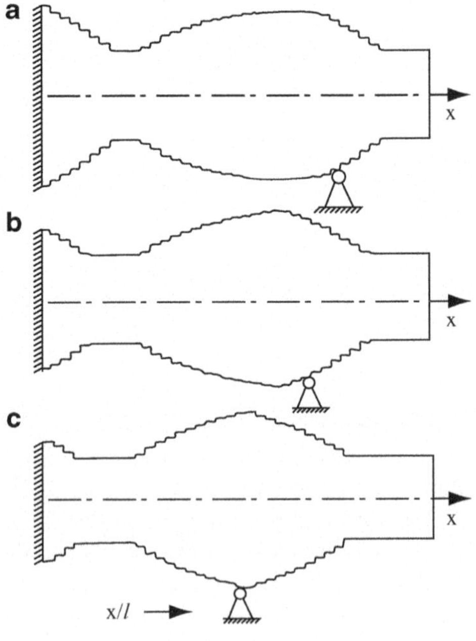

Fig. 3.59 Optimal supported microcantilevers at prescribed first natural frequency. *Top*: when $x_0/l = 0.7$. *Middle*: when $x_0/l = 0.6$. *Bottom*: when $x_0/l = 0.5$

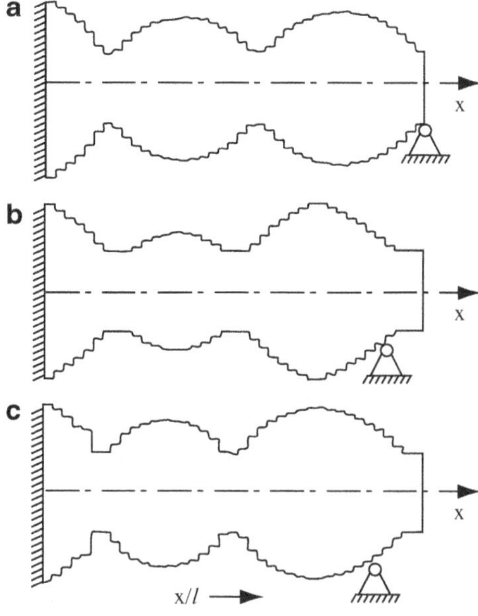

Fig. 3.60 Optimal supported microcantilevers at prescribed second natural frequency. *Top*: when $x_0/l = 1.0$. *Middle*: when x_0/l=0.9. *Bottom*: when $x_0/l = 0.8$

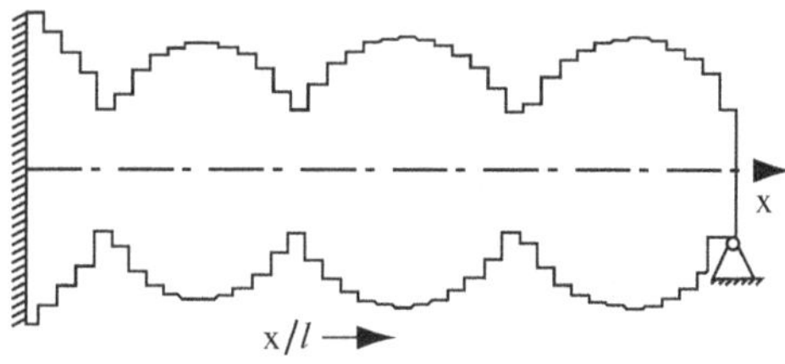

Fig. 3.61 Optimal supported microcantilever at the frequency ω_3 when $x_0/l = 1.0$

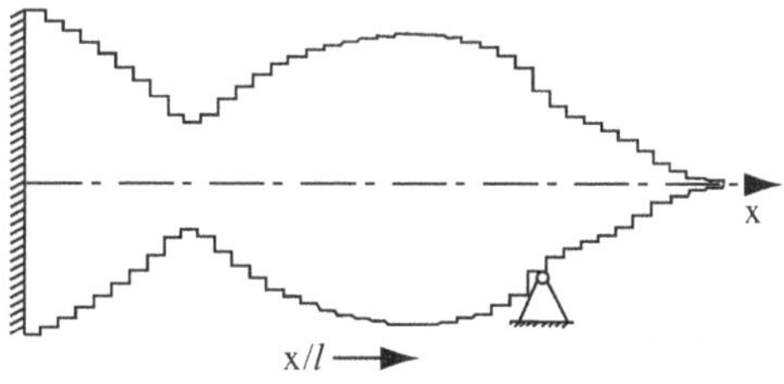

Fig. 3.62 Optimal supported microcantilever at frequency $\omega_2/\omega_1 = 1.3$ when $x_0/l = 0.75$

cross-sections at the distances $x_0/l = 0.2$ and $x_0/l = 0.5$ in respect of the support location site. The location of both minimum and maximum cross-sections in optimal supported and unsupported microcantilevers correspond to the particular points of the cross-section torsional vibration modes at a prescribed frequency of design.

The optimal structure presented in Fig. 3.61 was obtained for the third natural frequency when the support is in point $x_0/l = 1.0$. The optimal structure presented in Fig. 3.62

was obtained when the support is in point $x_0/l = 0.75$ and when the constraints of the state variables (frequencies ω_1 and ω_2), lead to a-minimum difference between the first ω_1 and the second ω_2 frequencies. Therefore, the ratio of the natural frequencies of the structure is $\omega_2/\omega_1 = 1.3$.

3.9.2 Optimization Results for Structures Undergoing Transient Vibrations

Requirements for most VIMs include the necessity of their adequate description by certain mathematical models, and the analysis of such models is to be performed by the iteration optimal design procedure. The method of model analysis must correspond to the design iteration cycle and be based on its advantages. In the solution of the mathematical programming task, it is necessary to establish the sensitivity of the dynamic characteristics of the model to the design sensitivity.

In the design of VIM, the following criteria are taken into account: the maximum rebound amplitudes of the impact-making links should not exceed certain values; the natural frequencies of the structure should not fall into certain frequency ranges. These criteria enable to avoid spontaneous connection or disconnection of links due to the effect of external impact or excitation force.

Here we provide an algorithm of optimal design of VIM meshed by finite elements. VIM is modeled by using Newmark's linear integration method, and optimal design is performed by the gradient projection method.

The system is described by the vector of design variables $\{A\} = \{A_1, A_2, \ldots, A_m\}^T$ where A_i – the areas of the finite element cross-sections. The task of optimal design is to choose the vector of design variables. The finite element model of the optimally designed VIM is provided in Fig. 3.1 VIM should be described in such a way that the VIM were optimal in a certain sense and its state characteristics would satisfy the preset constraints.

To specify the constraints on the VIM dynamic characteristics, the vector of the state variables $(y(t)\} = \{y_1(t), \ldots, y_k(t)\}^T$, is introduced for describing generalized displacement, speed and acceleration. The quality criterion depends on both the design variables and state variables.

In an appropriate choice of generalized displacements in the inertial system of coordinates, the state of the VIM may be described by the matrix equations of the following form:

$$[M(A)]\{\ddot{y}(t)\} + [C(A)\{\dot{y}(t)\} + [K(A)\{y(t)\} = \{Q(T)\} + \{Fy(t), \dot{y}(t), t\}, \quad (3.41)$$

with the initial conditions

$$\{\dot{y}(0)\} = \{\dot{y}^0\}; \ \{y(0)\} = \{y^0\}. \quad (3.42)$$

The scale of all matrices and vectors is equal to the number of degrees of freedom of the system.

During the elastic link impact against the support, the state Eq. (3.41) gains the following form:

$$\begin{aligned}&[M(A)]\{\ddot{y}(t)\} + [C(A) + \bar{C}]\{\dot{y}(t)\} + [K(A) + \bar{K}]\{y(t)\}\\&\quad = \{Q(T)\} + [\bar{K}]\Delta\},\end{aligned} \tag{3.43}$$

where $[\bar{K}]$ – the diagonal stiffness matrix of the support, $[\bar{C}]$ – the diagonal matrix of viscous friction coefficients of the support, $\{\Delta\}$ – the vector of the distances of the support from the microcantilever in equilibrium.

Similar as in the case of optimal design with respect to periodic vibration laws, the target functional is the structure mass:

$$\Phi_0(A) = g(A). \tag{3.44}$$

As during the analyzed interval of time the displacement of the link is required to vary within preset limits, we determine the constraint:

$$\{y(t)\} - \{y_{max}\} \leqslant 0, \quad \text{when } t \in [0, T] \tag{3.45}$$

where $\{y_{max}\}$ – the vector of displacement constraints; [0,T] the assigned interval of time.

Constraints of type (3.45) may be rewritten as $\{h(y(t), A, t)\} - \{y_{max}\} \leqslant 0$, when $t \in [0, T]$,and transformed to the integral constraints

$$\{\Phi_1(y(t), A, t)\} = \int_0^T \{h_j(y(t), A, t)\}dt = 0, \tag{3.46}$$

$$h_j[y(t), A, t] = \begin{cases} h_j[y(t), A, t] \\ 0. \end{cases}, \quad if \begin{matrix} h_j[y(t), A, t] > 0 \\ h_j[y(t), A, t] < 0 \end{matrix}. \qquad j = 1, 2, \ldots, s$$

Constraints imposed on design variables are expressed by the inequality:

$$A \leqslant A_j \leqslant \bar{A}, j = 1, 2, \ldots, m, \tag{3.47}$$

where A and $\bar{A}$ – vectors of the lowest and highest values of design variables.

As the constraints of (3.47) do not depend on time, they may be expressed in the following

$$\{\Phi_1(A)\} \leqslant 0.$$

In each iteration of the optimization process, either one or several constraints superimposed on the characteristics of the designed system may become active. These constraints are considered when determining the variation of the design variable. By denoting the active integral constraints in (3.46) by $\{\tilde{\Phi}_1[y(t), A]$and the active constraints in (3.47) by $\{\tilde{\Phi}_2(A)\}$, the vector of active constraints may be set up as

$$\{\tilde{\Phi}[y(t), A] = \{q\{A)\} + \int_0^T \{p[y(t), A, t]\}dt,$$

where

$$\{q\{A)\} = \left\{ \begin{matrix} 0 \\ \tilde{\Phi}_2(A) \end{matrix} \right\}; \int_0^T \{p[y(t), A, t]\}dt = \left\{ \begin{matrix} \tilde{\Phi}_1[y(t), A \\ 0 \end{matrix} \right\}. \tag{3.48}$$

Then the functional becomes

$$P\,[y(t), A] = G(A) + \int_0^T B[y(t), A, t]dt. \tag{3.49}$$

The first variation of the quality functional (3.44) is equal to

$$\delta\Phi_0 = \left\{ \frac{\partial g(A)}{\partial\{A\}} \right\} \delta\{A\},$$

and the first variation of constraints (3.49) becomes:

$$\delta P = \left\{ \frac{\partial G(A)}{\partial A} \right\} \delta\{A\} + \int_0^T \left(\left\{ \frac{\partial B[y(t), A, t]}{\partial\{y(t)\}} \right\} \delta\{y(t)\} + \left\{ \frac{\partial B[y(t), A, t]}{\partial\{A\}} \right\} \delta\{A\} \right) dt. \tag{3.50}$$

In order to eliminate $\delta\{y(t)\}$ in (3.50), it is necessary to calculate the first variation of both sides of the Eq. (3.43)

$$\begin{aligned} &[M(A)]\{\ddot{y}(t)\} + [C(A)]\delta\{\dot{y}(t)\} + [K(A)\delta\{y(t)\} \\ &= -\left(\left[\frac{\partial[M(A)]\{\ddot{y}(t)\}}{\partial\{A\}} \right] + \left[\frac{\partial[C(A)]\{\dot{y}(t)\}}{\partial\{A\}} \right] + \left[\frac{\partial[K(A)]\{y(t)\}}{\partial\{A\}} \right] \right) \delta\{A\}dt, \end{aligned} \tag{3.51}$$

$$\delta\dot{y}(0) = 0; \delta y(0) = 0.$$

Denoting the right side of Eq. (3.51) by $\left[\frac{\partial\{R(A,t)\}}{\partial\{A\}}\right]\delta\{A\}$ and multiplying it by the transposed conjugated vector $\{\lambda(t)^T\}$, in the integral form, the following expression is obtained:

$$\begin{aligned}&\int_0^T \{\lambda(t)\}^T([M(A)]\{\ddot{y}(t)\} + [C(A)]\delta\{\dot{y}(t)\} + [K(A)\delta\{y(t)\})dt\\ &= \int_0^T \{\lambda(t)\}^T\left[\frac{\partial\{R(A,t)\}}{\partial\{A\}}\right]\delta\{A\}dt.\end{aligned} \tag{3.52}$$

The expression (3.52) is integrated during the analyzed interval of time

$$\begin{aligned}&\{\lambda(t)\}^T[M(A)]\delta\{\dot{y}(t)\}\big|_0^T - \{\dot{\lambda}(t)\}^T[M(A)]\delta\{y(t)\}\big|_0^T + \{\lambda(t)\}^T[C(A)]\delta\{y(t)\}\big|_0^T\\ &+ \int_0^T \left[\{\ddot{\lambda}(t)\}^T[M(A)]\delta\{y(t)\} - \{\dot{\lambda}(t)\}^T[C(A)]\delta\{y(t)\} + \{\lambda(t)\}^T[K(A)]\delta\{\dot{y}(t)\}\right]dt\\ &= \int_0^T \{\lambda(t)\}^T\left[\frac{\partial\{R(A,t)\}}{\partial\{A\}}\right]\delta\{A\}dt.\end{aligned} \tag{3.53}$$

Considering (3.51) to eliminate non-integral quantities determined at the final time moment from (3.53), it should be required that

$$\begin{aligned}&\{\lambda(T)\}^T[M(A)]\delta\{\dot{y}(T)\} - \{\dot{\lambda}(T)\}^T[M(A)]\delta\{y(T)\}+\\ &\{\lambda(T)\}^T[C(A)]\delta\{y(T)\} = 0.\end{aligned} \tag{3.54}$$

Since variations $\delta\dot{y}\{(T)\}$and $\delta y\{(T)\}$ are arbitrary, and $[M(A)]$ and $[C(A)]$ are non-singular, (3.54) will be satisfied, provided $\{\dot{\lambda}(0)\} = \{\lambda(0)\} = 0$. The conjugate vector is determined as the solution of the equation

$$[M(A)]\{\ddot{\lambda}(t)\} - [C(A)]\{\dot{\lambda}(t)\} + [K(A)]\{\lambda(t)\} = \left\{\frac{\partial B[y(t), A, t]}{\partial\{A\}}\right\}^T, \tag{3.55}$$

derived from (3.43) and (3.50).

Introducing $t = T\text{-}\eta$, $\{\lambda(t)\} = \{z(\eta)\}$ and $B[y(t), A, t] = B'[y(\eta), A, \eta]$, the Eq. (3.55) and the condition $\{\lambda(t)\}$ at the final time moment are transformed to the form:

$$[M(A)]\{\ddot{z}(\eta)\} - [C(A)]\{\dot{z}(\eta)\} + [K(A)]\{z(\eta)\} = \left\{\frac{\partial B'[y(\eta), A, \eta]}{\partial\{y(\eta)\}}\right\}^T; \quad (3.56)$$

$$\{\dot{z}(0)\} = \{z(0)\} = 0.$$

The Eq. (3.56) corresponds to the Eq. (3.43). Equation (3.56) is solved by the Newmark direct integration method.

Determining $\{z(\eta)\}$ and $\{\lambda(t)\}$, the right side of the Eq. (3.53) may be substituted into

$$\{\dot{z}(0)\} = \{z(0)\} = 0 \quad (3.57)$$

instead of $\int_0^T \frac{\partial B'[y(\eta),A,\eta]}{\partial\{y(\eta)\}}\delta\{y(t)\}dt$, and the first variation δP may be expressed by term of $\delta\{A\}$:

$$\delta P = \left(\left\{\frac{\partial G(A)}{\partial A}\right\} + \int_0^T \left(\left\{\frac{\partial B[y(t), A, t]}{\partial\{y(t)\}}\right\} + \{\lambda(t)\}^T\left\{\frac{\partial\{R(A,t)\}}{\partial\{A\}}\right\}\right)\right)\delta\{A\}.$$

By using this algorithm the computer program has been developed and optimal design of the VIM consisting of an elastic microcantilever and a rigid support has been performed. The microcantilever is set into motion by the initial deformation $y(t) = y^0$ and by releasing from the deformed position at the vibration moment $t = 0$. Having been released, the microcantilever is affected by support and having gained speed at time moment t_1, collides with it, rebounds from it and performs free damped impact vibrations as shown in Fig. 3.63. The problem was solved by determining the microcantilever shape that would maintain the assigned maximum amplitude y_{max} after the first impact. The problem was solved for different positions of the support. Modeling of the system behavior was carried out in the interval (t_1, t_2) between the first and second impact.

The calculated optimal configurations are presented in Figs. 3.64 and 3.65, for the support located at the points $x_0/l = 0$; 0.8; 0.9 and 1.0. In each case the microcantilever mass decreased by 20–30% as compared to the mass of the initial structure of constant cross-section possessing corresponding dynamic characteristics. Though the static characteristics of the obtained optimal structures are different, dissipation of energy in the material is enhanced, therefore the post-impact rebound amplitude decreases and thus the transient vibro-impact motion becomes shorter.

In addition to calculations of free impact vibration, optimal shapes of microcantilevers were obtained maintaining the prescribed rebound amplitudes, possessing shorter transient vibrations when the free end of the microcantilevers is affected by discrete excitation pulses. Such configurations for different positions of the support are presented in Figs. 3.66 and 3.67.

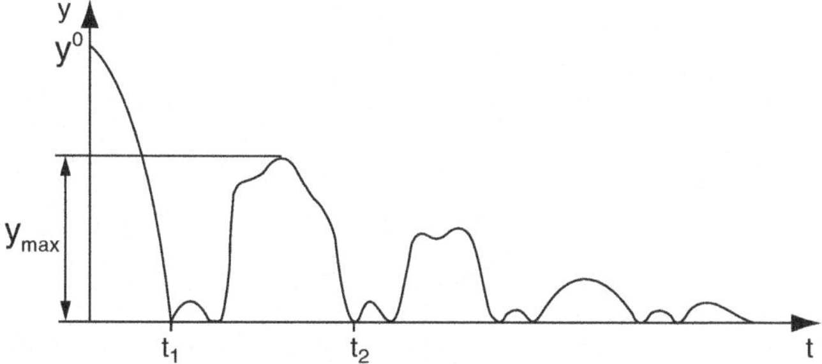

Fig. 3.63 The microcantilever motion trajectory in case of free damped impact vibration

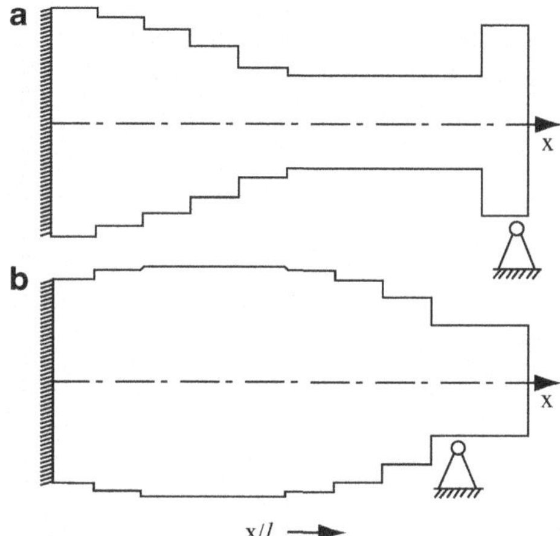

Fig. 3.64 Optimal microcantilevers at a prescribed rebound amplitude. *Top*: when $x_0/l = 1.0$. *Bottom*: when $x_0/l = 0.9$

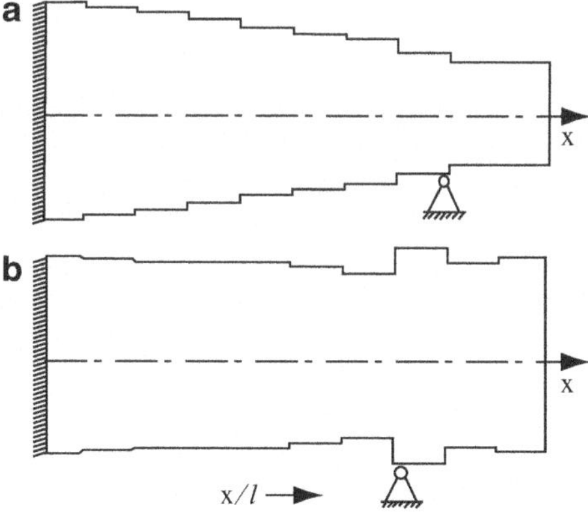

Fig. 3.65 Optimal microcantilevers at a prescribed rebound amplitude. *Top*: when $x_0/l = 0.8$. *Bottom*: when $x_0/l = 0.7$

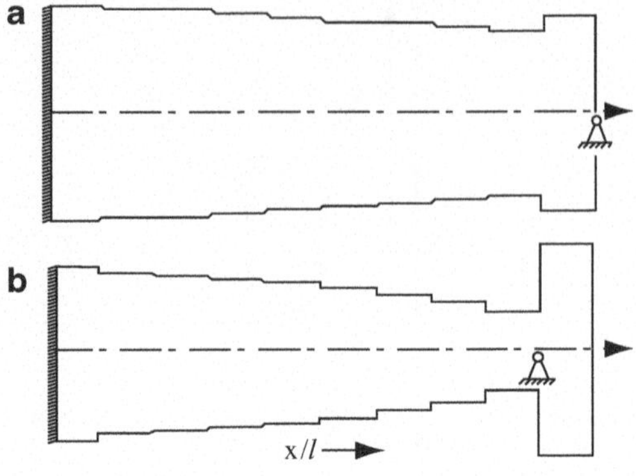

Fig. 3.66 Optimal microcantilevers at a prescribed rebound amplitude in the case of discrete excitation. *Top*: when $x_0/l = 1.0$. *Bottom*: when $x_0/l = 0.9$

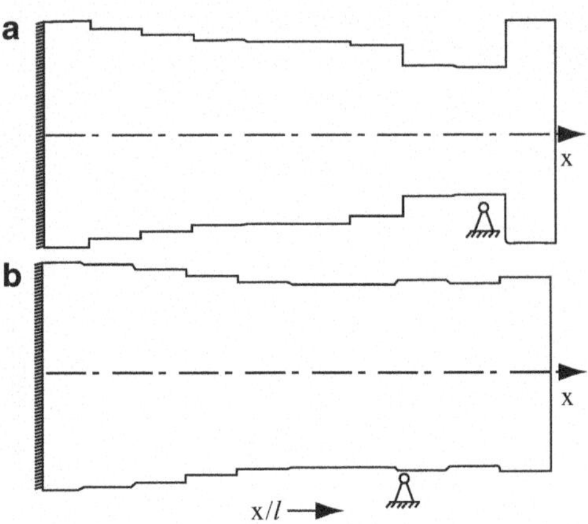

Fig. 3.67 Optimal microcantilever at a prescribed rebound amplitude in the case of discrete excitation. *Top*: when $x_0/l = 0.8$. *Bottom*: when $x_0/l=0.7$

One of the most important requirements for the VIMs is their vibrational and impact stability. One of possible ways to increase it is by choosing a certain shape of the elastic link that would ensure smallest amplitudes of rebound from the support in case of kinematic excitation. Upon relevant choice of generalized displacements in the inertia system of coordinates, the kinematically excited structure can be described by the equation:

$$\begin{aligned}&[M_{SS}(A)]\{\ddot{y}_S(t)\} + [C_{SS}(A)]\{\dot{y}_S(t)\} + [K_{SS}(A)]\{y_S(t)\}\\&= \{Q_S(t)\} + \{F((y_S(t), \dot{y}_S(t), t)\} - [M_{SK}(A)]\{\ddot{y}_K(t)\}\\&+ [C_{SK}(A)]\{\dot{y}_K(t)\} + [K_{SK}(A)]\{y_K(t)\}\end{aligned}$$

and the initial conditions

$$\{\dot{y}_S(0)\} = \{\dot{y}_S{}^0\};\ \{y_S(0)\} = \{y_S{}^0\}.$$

Fig. 3.68 Optimal microcantilever at a prescribed rebound amplitude in the case of kinematic sine excitation

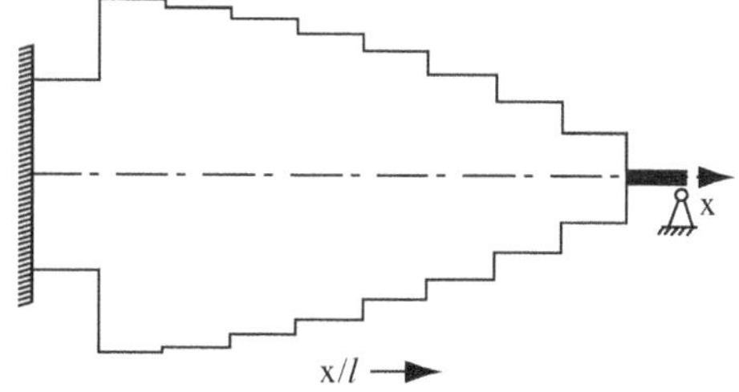

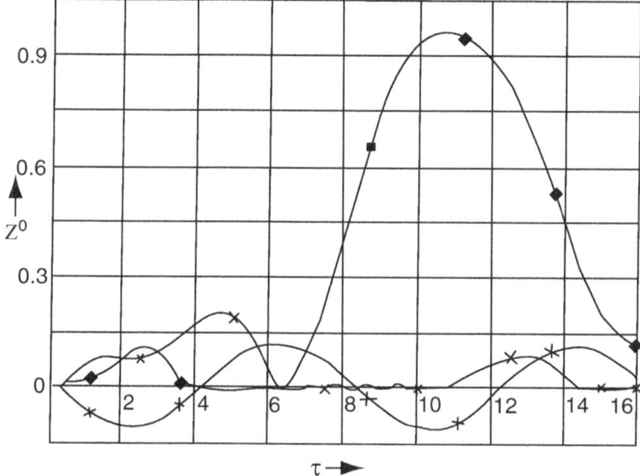

Fig. 3.69 The vibration law of the microcantilever in the case of kinematic excitation: ♦– initial structure of constant cross-section, x – optimal structure

Figure 3.68 provides an optimal structure for harmonic sine kinematic excitation, when the frequency exceeds the first natural frequency of the microcantilever by three times ($\omega = 3\omega_1$). In this case the support is located at the free end of the microcantilever, and the prestress is equal to zero.

The motion laws provided in Fig. 3.69 both for the initial and optimal structure (Fig. 3.68) show that the impact vibration amplitudes of the optimal structure is almost by five times smaller than the amplitudes of the initial structure. In the case when the stress is obtained by displacing the support towards the microcantilever structure by a distance equal to the rebound amplitude, the VIM in which the links are normally connected, is vibrationally stable at prescribed parameters of kinematic excitation.

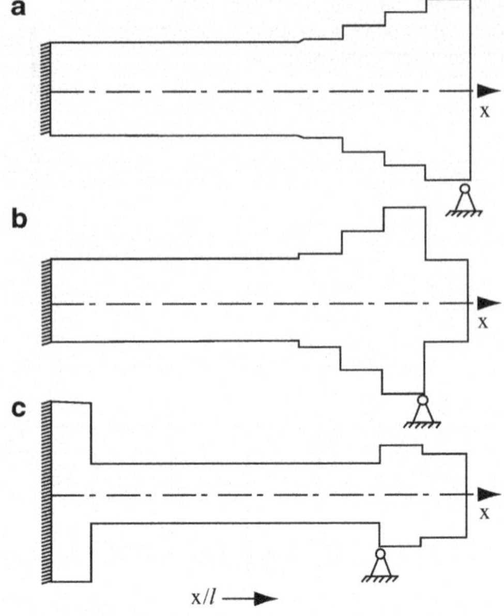

Fig. 3.70 Optimal microcantilevers at kinematic impact excitation when the pulses are directed from the support. *Top*: when $x_0/l = 1.0$. *Middle*: when $x_0/l = 0.9$. *Bottom*: when $x_0/l = 0.8$

In addition to requirements for vibrational stability, requirements for impact stability are often imposed on VIMs. For the purpose, optimal structures were obtained that are stable with regard to kinematic impact excitation. Optimal configurations for the case when the kinematic excitation pulses are directed to have the elastic structure lifted from the support are presented in Figs. 3.70 and 3.71 for different positions of the support. The provided optimal shapes have one minimum cross-section which is comparable to the optimal structures described in the previous chapter for the second natural frequency of the transverse mode. This proves that when changing the shape of the elastic structure, vibrational and impact stability is secured by amplifying the amplitude of the second mode of transverse vibrations.

Figures 3.72 and 3.73 provide optimal configurations ensuring the amplitudes prescribed in the design process in the case of kinematic impact excitation, when the operating excitation pulses press the elastic link to the support.

Contrary to the previous case, the elastic link, affected by inertia forces, at the beginning lifts off the support, and then makes impacts against it. Such optimal structures are applicable for VIMs with a gap in between the impact links. As can be seen from Figs. 3.72 and 3.73, the obtained structures, irrespective of the position of the support, are different from the structures provided in Figs. 3.70 and 3.71, as they have two minimum values in the cross-section. This is typical for optimal structures

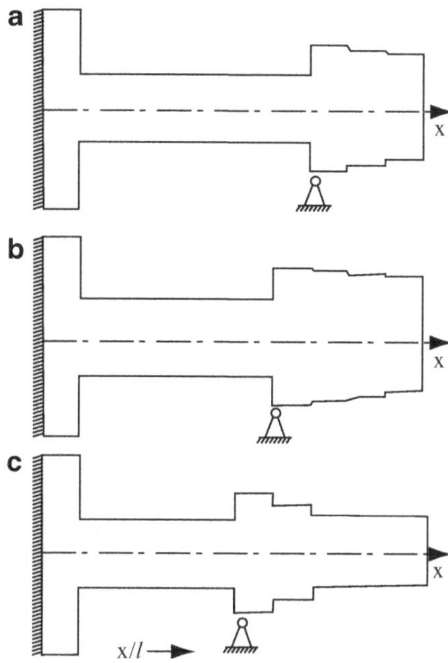

Fig. 3.71 Optimal microcantilevers at kinematic impact excitation when the pulses are directed from the support. *Top*: $x_0/l = 0.7$. *Middle*: $x_0/l = 0.6$. *Bottom*: $x_0/l = 0.5$

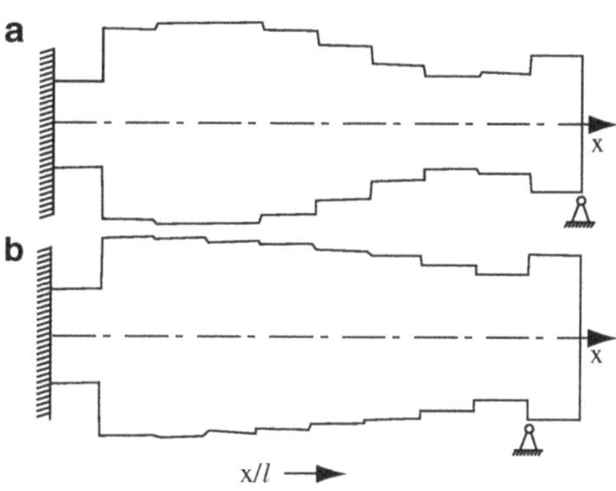

Fig. 3.72 Optimal microcantilevers at kinematic impact excitation when the pulses are directed to the support

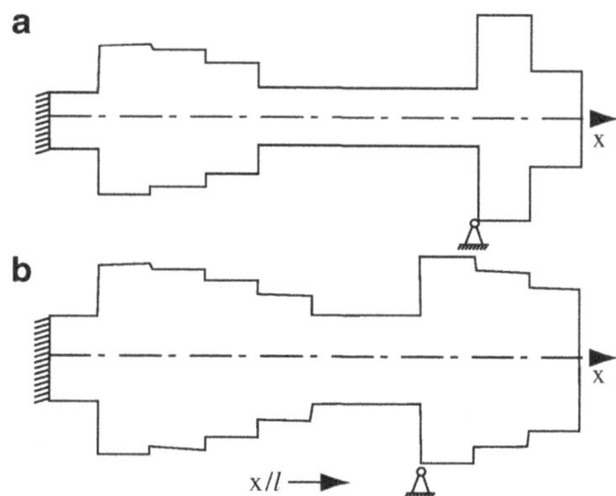

Fig. 3.73 Optimal microcantilevers at kinematic impact excitation when the pulses are directed to the support

obtained for the third natural frequency. It follows that the impact stability in VIMs with a gap between the impact making elements may be enhanced by intensifying the vibration amplitudes of the third mode.

References

1. Bathe KJ, Wilson EL (1976) Numerical methods in finite element analysis. Prentice-Hall, Englewood Cliffs, NJ
2. Ostasevicius V (1998) Dynamics and modeling of mechanical structures. Technologija, Kaunas, Lithuania, Lithuanian edition

Chapter 4
Theoretical Analysis of a Micromotor

Abstract This chapter is dedicated to analytical and numerical analysis of electrostatic micromotor with focus on key aspects of its design and control. FE modeling and simulation is reported providing results of vibration analysis of the rotor in vacuum and viscous medium. Analytical dynamic micromotor model is presented together with results of analysis of electrostatic forces between a pair of poles together with simulations of different motor configurations and investigation of torque curves. This chapter also describes the developed control methodology of pole switching sequences of MEMS motors.

Simulation of micromotors and other MEMS devices involves several physical effects such as mechanical motion, thermodynamics, electrostatics, electromagnetism, etc. Detailed knowledge of all these effects is a prerequisite for effective and efficient simulations. Since MEMS research is built on multidisciplinary foundations, a particular challenge here for MEMS is the development of effective analytical or numerical tools that deal efficiently with the modeling and simulation suites that are appropriate to its computational analysis requirements. Especially simulation is a very important technique for fast and low-cost analysis of different versions of a design and the estimation of system behavior.

In order to design such a micromotor that would ensure the required performance characteristics, it is vitally important to be able to investigate and evaluate the effects of its various geometric and material parameters on these characteristics. Traditionally, this is done either analytically or experimentally. Analytical approaches are usually quick but they are almost always applicable only to 2-D problems and designs with simple topology. In contrast, experimental techniques are apparently useful in many cases but their use is extremely expensive and very time-consuming. Considering this, with the increasing power of modern computer hardware and software, numerical techniques such as FE method seem to be a better alternative. The first reason to use FE calculation is that it enables modeling of different surface force laws and quantify how they affect structures. The second and ultimately more important reason for the FE calculations is that analytical solutions cannot be used to model the electrostatic loading in the actual experiment and are unable to easily include important effects such as fringing fields.

V. Ostasevicius and R. Dauksevicius, *Microsystems Dynamics*,
Intelligent Systems, Control and Automation: Science and Engineering 44,
DOI 10.1007/978-90-481-9701-9_4,

The extensive use of FE technique for solving partial differential equations of electrostatics in complex geometries and material nonlinearities makes it particularly suitable for actuator modeling and adaptable to computer-aided (CAD) design optimization and performance evaluation. The purpose of the model development is to understand, comprehend and utilize different phenomena and effects.

4.1 Finite Element Modeling Procedure

While the motor is in operation, the rotor is intended to be in electrical contact with the ground plane, and the rotor and axle form a pair of contact bodies. These lead to the conclusion that the rotor dynamics are underdamped, and are insufficient to develop a detailed model for the full dynamic behavior. In the micromotors, there is intermittent contact at the rotor–stator interface and physical contact at the rotor–axle interface. The particles appearing irregularly in the gap between the rotor and stator of electrostatic micromotor make the rotating state unpredictable (Fig. 4.1). Also the modeling of MEMS behavior requires a precise knowledge of the process parameters such as thickness, Young modulus, stress, roughness, etc. These technological parameters are generally very dependent of the deposition process. For instance, it has been observed that Young modulus can vary by a factor 2 and stress by a factor 100 regarding their deposition method [1]. This complicates micromotor model even more.

Motion becomes very difficult to determine analytically as structures become arbitrarily complex. For this reason, in FE analysis, the structure to be analyzed is discretized into small elements, each having an associated stiffness matrix which is stored mathematically in a lookup table, in the form of fundamental equations. When problem-specific parameters such as dimensional coordinates, the material elastic modulus, Poisson's ratio and density are put in these equations, the local stiffness, as represented by one element, is uniquely known. When a structure is fully discretized, or meshed, into many such elements, its global stiffness can be assembled, again in the form of a matrix, from the combined stiffnesses of all the interacting elements. If a force or set of forces is subsequently applied to the structure, the displacement response can then be calculated by inverting the global

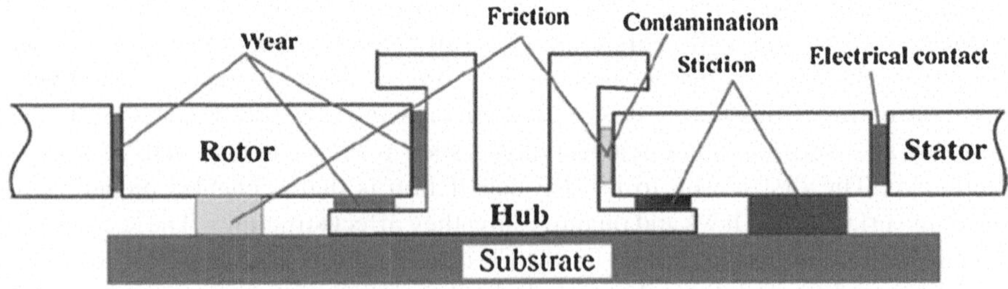

Fig. 4.1 Schematic of an electrostatic micromotor

stiffness matrix $\{x\} = \{F\}\{k\}^{-1}$ (where $\{F\}$ – applied force vector, $\{k\}$ – stiffness matrix and $\{x\}$ – displacement vector.

This basic concept can be used in the solution of many problems involving a variety of applied loading conditions, including externally applied static forces, dynamic forces, electric fields, pressures and temperatures.

Simulating complete behavior of most microstructures requires solving not just for quantities in the interior of the structure (such as displacement) but also for exterior quantities (such as electric fields). Different kinds of calculations cannot be performed always with a single simulator. Computational analysis of electrostatically actuated MEMS requires a self-consistent solution of the coupled mechanical and electrical equations. When a voltage is applied between a rotor and a stator, electrostatic charges are induced (Fig. 4.2). These charges give rise to electrostatic forces, which act normal to the surface of the conductors. Since the stator is fixed and cannot move, the electrostatic forces move only the rotor. When the rotor moves, the charge redistributes on the surface and, consequently, the resultant electrostatic forces and the position of poles also change. This process continues until an equilibrium state is reached. These mechanical and electrostatic problems must be solved self-consistently. Simply computing the electrostatic forces on the undeformed/unmoved structure may no be accurate. In general, motion or mechanical deformation of the structure will cause charge redistribution, and the electrostatic forces on the deformed structure will differ from those on undeformed structure. Thus, mechanical analysis is performed to compute the position of the structures according to given electrostatic forces. Electrostatic analysis is

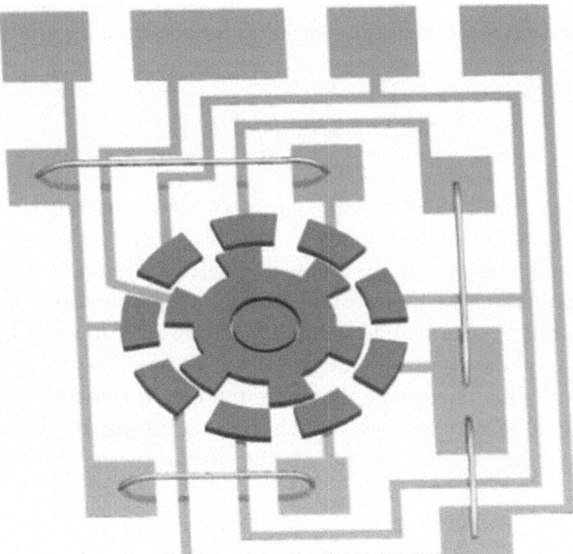

Fig. 4.2 3-D schematic of a developed electrostatic micromotor with electric connection pads and leads indicated

performed to compute the surface charge densities and the electrostatic forces. Mechanical analysis is typically implemented using the initial or the undeformed configuration. When the structural displacements are computed using the mechanical analysis, the geometry of the structures is updated and electrostatic analysis is performed on the deformed configuration to compute the surface charge densities. Once the surface charge density is known, the electrostatic forces are computed on the deformed configuration and transformed onto the initial configuration. Using the computed electrostatic forces a mechanical analysis is again performed in the initial configuration to recompute the structural displacements. This procedure is repeated until an equilibrium position is reached.

A major difficulty at each iteration is the need to update geometry of the structures before electrostatic analysis is performed. This presents several problems:

1. Flat surfaces of the structures in the initial configuration can become curved due to deformation.
2. When the structure undergoes very large deformation, remeshing the surface may become necessary before an electrostatic analysis is performed.
3. Interpolation functions, used in many numerical methods, need to be recomputed whenever the geometry changes.

Each of these issues significantly increases the computational effort making the self-consistent analysis of electrostatic MEMS an extremely complex and challenging task [2].

In order to obtain the electrostatic torque as a function of rotor position, the electrostatic energy must first be calculated. The latter can be found by performing field analysis for each rotor position. At each rotor position, the stored energy in the electric field is evaluated as a function of rotor position:

$$W(\phi) = \frac{1}{2}\varepsilon_0 \iiint E^2 dV, \tag{4.1}$$

where W – energy of the electric field, ε_0 – free space permittivity, φ – angle of rotation, E – electrostatic field intensity, V – potential difference between electrodes.

Each new rotor position requires a new mesh generation. These positions are all considered in evaluating the electrostatic energy of the micromotor. After a set of energy–angle points is obtained, a continuous curve is fit to them by interpolation techniques, and then the energy vs. angle curve is found.

After calculating the energy–angle curve, the electrostatic torque is then found by using the principle of virtual work, wherein the torque is given by the partial derivative of the stored energy vs. the angular displacement, φ, of the rotor

$$T = \frac{\partial W}{\partial \phi} = \frac{1}{2} V^2 \frac{\partial C}{\partial \phi}, \tag{4.2}$$

where C – capacitance.

Since the electrostatic micromotor is an electrically linear system, the driving torque is calculated per meter of axial thickness and can then be multiplied by the actual axial thickness.

4.2 Scaling in MEMS

When the dimensions of a macroscopic object shrink to micro-scale, the surface area to volume ratio increase and component masses and the inertial force of the object is no more a dominant factor. At the same time, the interfacial forces, such as the capillary force and electrostatic forces emerge and become dominant. Friction and wear in micro-scale is inherently related to the surface forces. Tribological designs of micromotors need to take into account that surfaces will be subjected to friction, wear and geometry variation [3].

There are a few matters one needs to keep in mind when miniature technical systems are designed. Linear dimensions shrink slower than surfaces and surfaces shrink slower than volumes. So, friction associated with surfaces becomes relatively more important than masses. But also stiffness, electric forces, magnetic forces, adhesive forces as well as the character of convection, surface finish and characteristic times associated with diffusion and resonance all scale in their own way. It is thus instructive to analyze scaling by comparing forces and effects by studying the exponent of the length scale of linear dimensions l. Then, linear dimensions l scale as $l \sim l^1$, surfaces S scale as $S \sim l^2$ and volumes V scale as $V \sim l^3$. So when a system is reduced isomorphically in size, the ratio of surface area to volume increases, rendering the, surface forces more important. Because shrinking the linear dimensions does not shrink forces in the same way, physical properties scale in various ways that are summarized in Table 4.1 [4, 5].

One scaling law that is helpful for micro actuators is the increased break down field strength of very small gaps due to the Paschen's law [5, 6] Electrical breakdown for devices with gaps smaller than 5 μm is different than in macro world because breakdown voltage is a function of the reduced variable of the pressure-gap spacing product. For gaps greater than about 10 μm, breakdown occurs when the electric field exceeds about 3 V/μm. For gap spacing smaller than 5 μm the field strength is on the order of 75 V/μm (Fig. 4.3). Assuming air at atmospheric pressure and gap spacing equal to 5 μm, the voltage at which the electrical breakdown occurs is about 350 V as opposed to ~20 V/μm as could be expected. Thus for small sizes, the obtainable electrostatic forces can be stronger than electromagnetic forces.

Other interesting effects that are safely ignored at macroscopic scales can become important in the microworld, including quantized thermal conductance, Brownian motion, mesoscopic mechanical properties and exotic dislocation line motions [7].

Table 4.1 Scaling of various physical properties

Parameter	Unit	Scaling
Linear dimension	l	$\sim l^1$
Surface	A	$\sim l^2$
Volume	V	$\sim l^3$
Electrostatic energy	W_{el}	$\sim l^3$
Electrostatic force	F_{el}	$\sim l^2$
Magnetic energy	W_{mag}	$\sim l^5$
Magnetic force	F_{mag}	$\sim l^4$
Structure deflection under own weight	ζ	$\sim l^2$
Stable length	L_{cr}	$\sim l^{4/3}$
Diffusion time	τ	$\sim l^2$
Drift velocity	v	$\sim l^2$
Transient time	τ	$\sim l^2$
Electric resistance	R_{el}	$\sim l^{-1}$
Hydraulic resistance	R_{hy}	$\sim l^{-3}$
Reynolds number	R_e	$\sim l^2$
Resonant frequency	f	$\sim l^{-1}$
Squeeze film damping force	F	$\sim l^1$
Power density		$\sim l^{-1}$

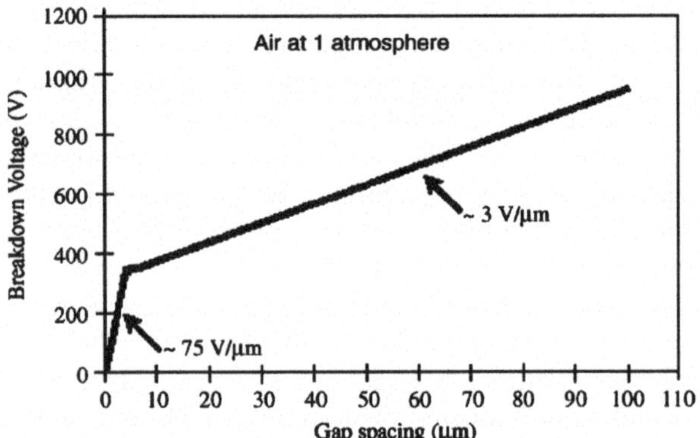

Fig. 4.3 Paschen curve: breakdown voltage vs. gap spacing at 1 atm

4.3 Modal Analysis of a Microrotor

Modal analysis is usually a first and basic step in analysis of numerical models of structures. Modal analysis gives displacement maxima in a vibration event and it reveals the frequencies of natural vibration, thus it can be used to predict the stochastic response of the device when it is excited by random vibration. It should be noted that these are naturally occurring resonant excitation states of the device, and will be excited if a harmonic forcing function is applied to the unit.

According to literature sources, speeds of micromotors can be up to 100,000 rpm, i.e. 16,000 rotations per second. If natural vibration modes are in this range, the motion of the rotor can become highly unpredictable or even destructive. For this reason modal analysis needs to be performed.

Center axle and stator poles of a motor are stably attached to wafer surface and cannot vibrate, whereas the rotor is of particular interest, because it can move freely. Rotors of electrostatic motors can be produced very different, both geometrically and using different materials. Depending on the way a rotor is produced, it can be thin (up to 30 μm, though mostly 1–10 μm) using UV photolithography or thicker using X-Ray lithography and electroplating. Due to the ratio area/thickness, the former ones match a typical definition of a plate. The latter ones match a definition of a solid body. Thus, the thicker rotors practically cannot be excited into detectable vibrations because of their geometric properties, but thin photolithographic films will vibrate. The following boundary conditions for modeling were determined to be: material – nickel, rotor radius 50, ..., 800 μm, thickness 1, ..., 20 μm, the number of poles, geometry of rotors was not restricted. Nickel was selected because this was one of the most common materials in MEMS manufacture and also this was the actual material used for producing of the motor. According to earlier reviewed articles, most popular rotor sizes vary around 100–300 μm, and for the following analysis only thin rotors are of interest, thus their thickness was restricted up to 20 μm.

A simple round plate can be excited to vibrate at many normal mode frequencies and their higher harmonics but their dependencies are simple and predictable. There are a number analytical formulas that can be readily applied. But analysis of a real microrotor having a hole in the center and poles on sides is much more complicated. In most cases FE analysis is necessary.

Modal analysis was firstly performed using finite element modeling program ANSYS. The first models created were fairly simple, assuming the rotor is a circular plate that was treated axisymmetrically. Solutions were accurate but as the method implies, only axisymmetric vibration modes were calculated. Thus, further improved 3-D models were created that repeated the results of axisymmetric calculations and added additional vibrations modes (Table 4.2). Out of the selection of numerically calculated modes, only the first five modes that would likely occur in were selected (Table 4.3). The sixth mode depends substantially on size and geometry of rotor poles and was selected only as a special case scenario. Theoretically, vibrations along rotor are also possible, but due to higher energy required, they are insignificant in practice.

Nodal lines in the graphs were represented by dots. Vibration modes of the rotor were classified into three groups: axisymmetric (mode 1), plane vibrations (modes 2–5) and twisting of poles (mode 6). All of them have higher harmonics, too. The first mode is independent on the number or geometry of poles, whereas the latter four are dependent to some degree.

Basic round plates with a hole were analyzed first. There is an obvious decrease in all normal mode frequencies with an increase of rotor size, if the hole is kept of a constant diameter (Fig. 4.4).

Table 4.2 Simulated vibration modes of a rotor

Vibration form	Mode (kHz)	Mode (kHz)	Mode (kHz)
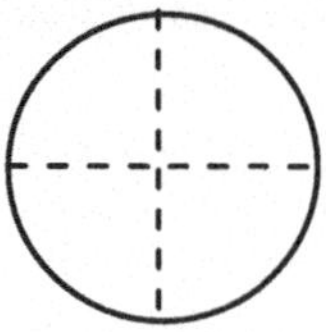	 385	 1,970	 3,490
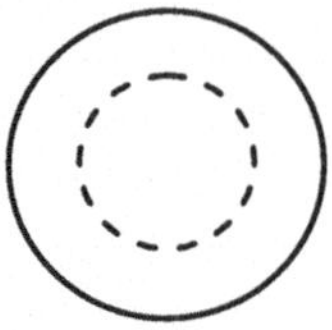	 663	 3,900	
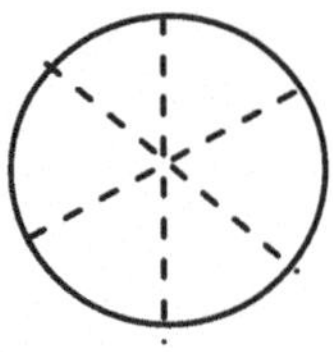	 771	 2,270	 4,000
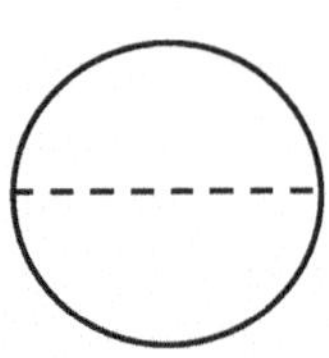	 1,320	 3,410	 4,540

(continued)

Table 4.2 (continued)

Vibration form	Mode (kHz)	Mode (kHz)	Mode (kHz)
	3,570		

Table 4.3 Basic vibration modes of a rotor

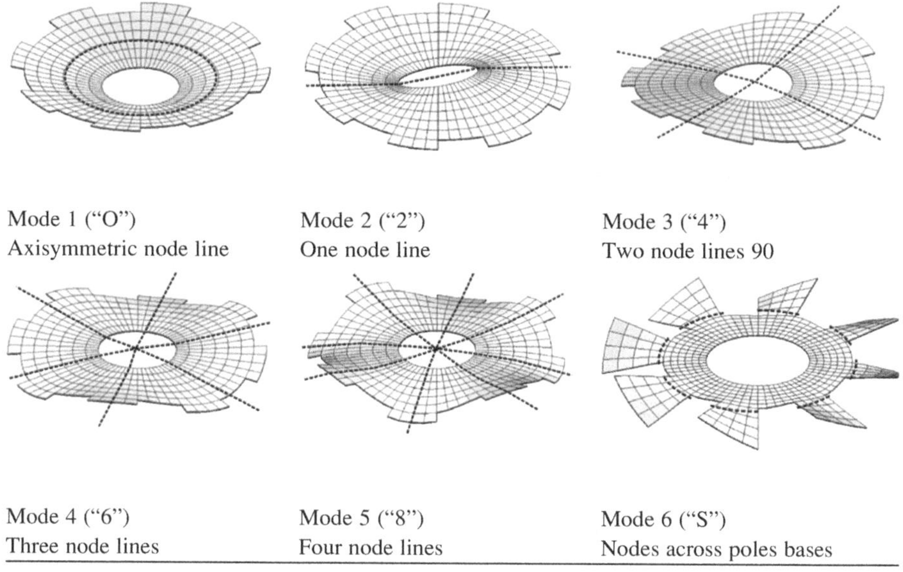

Mode 1 ("O") Axisymmetric node line	Mode 2 ("2") One node line	Mode 3 ("4") Two node lines 90
Mode 4 ("6") Three node lines	Mode 5 ("8") Four node lines	Mode 6 ("S") Nodes across poles bases

Similar results were obtained for round plates having a higher diameter hole, though a sharp increase in some lower frequencies was apparent for these rotors (Fig. 4.5). This happened because the outside diameter of the rotor was decreased close to the size of internal hole, so the overall proportions of the rotor changed sharply. Higher modes were not affected. This was an unexpected, but clear result.

Analysis of 8-pole rotor models showed nearly no change in normal mode frequencies and their distribution, as compared to simplified circular plates

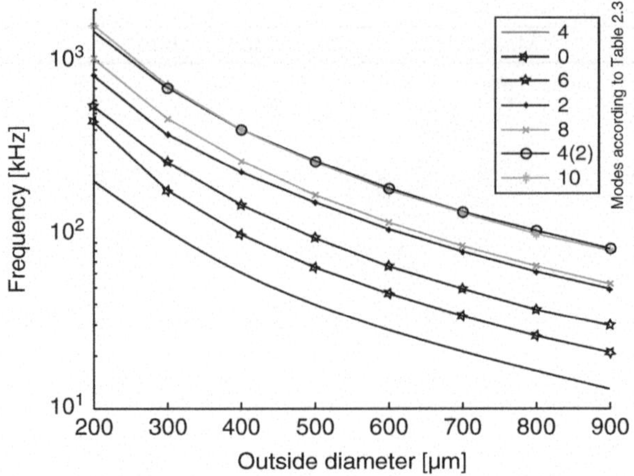

Fig. 4.4 Dependence of frequency of vibration modes on the size of circular plate with Ø100 μm axle hole

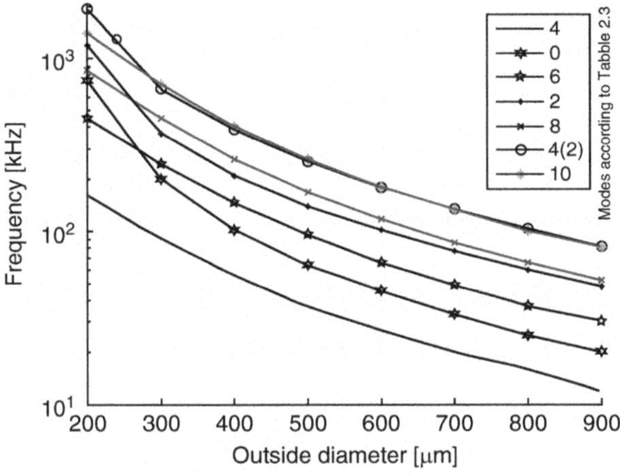

Fig. 4.5 Dependence of frequency of vibration modes on the size of circular plate with Ø150 μm axle hole

(Fig. 4.6). This suggested rather great possibilities in design of multi-pole rotors, though additional analysis was still necessary.

Further analysis was done studying the influence of number of poles on vibration modes. Both poles and the gaps between the poles were set equal. By keeping all parameters the same and changing only number of poles, the following results were obtained, as shown in Fig. 4.7. There was practically no change in frequencies of lower modes (1–4) as the number of poles was increased (Fig. 4.7 (top)). This happened mainly because the mass and volume of rotor remained constant. Modes 5 and 6 are highly dependent on geometry of the rotor, thus resulted in bigger

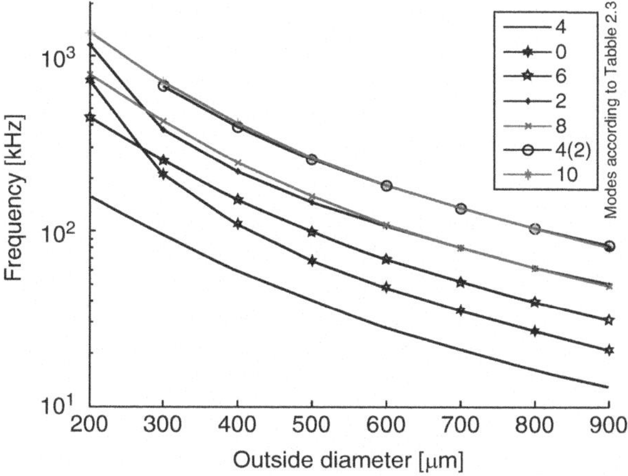

Fig. 4.6 Dependence of frequency of vibration modes on the size of circular plate with Ø150 μm axle hole having 8 poles (pole diameter – 10% of rotor radius)

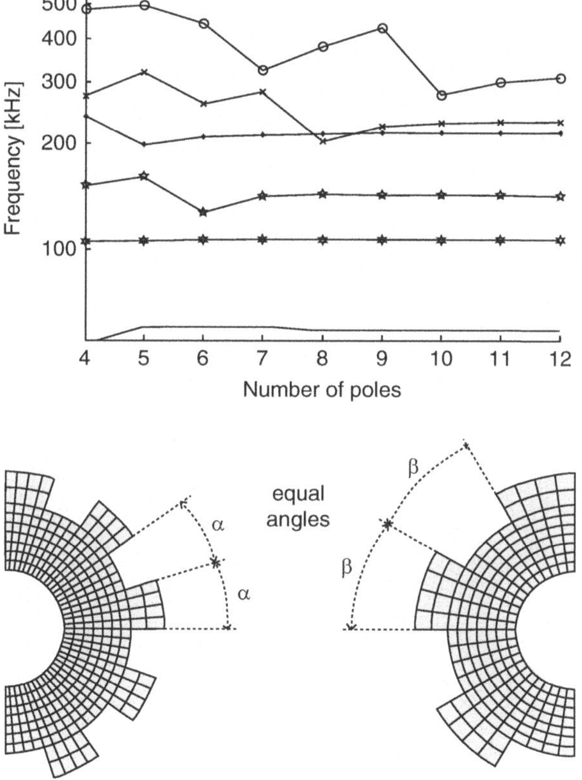

Fig. 4.7 Frequency of vibration modes depending on the number of poles (poles take up 50% of perimeter)

frequency shifts (Fig. 4.7 (bottom)). For instance, mode 5 is sharpest if rotor has even number of poles. Thus, rotors having an uneven number of poles are less prone to going into resonance in modes 2–5.

On the other hand, if poles are set to some particular size (constant angle) the results differ, though the frequencies of vibration modes are still very dependable on the number of poles the rotor has (Fig. 4.8 (top)). But if in the previous case increasing the number of poles resulted in practically no change in vibration frequencies, here on the contrary, frequencies of all modes decrease as the number of poles increases (Fig. 4.8 (bottom)). This happens because the mass of the rotor increases with every additional pole. So, ultimately it would result in becoming just a plane disc of a bigger diameter. These calculations confirm earlier results given in Figs. 4.4 and 4.7.

Rotor can be considered a circularly patterned body around its center axis. Results of calculations reveal that higher modes are very influenced by the multiple of pattern. Thus, vibrations in mode 3 having two nodal lines need less energy for rotors having a multiple of 4 poles, i.e. a total of 4, 8, 12, etc. poles; mode 4 (three

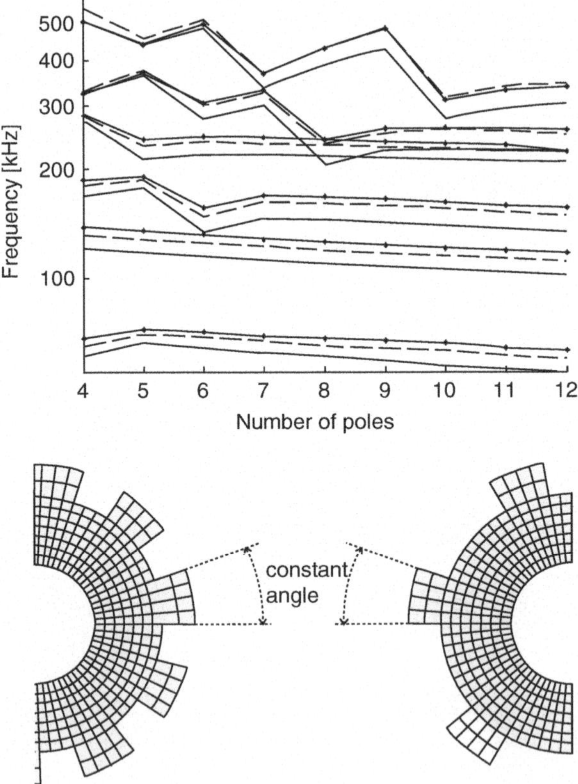

Fig. 4.8 Frequency of vibration modes depending on the number of poles. *Dotted line* – 10°, *continuous line* – 20°, *dashed line* – narrowing from 20° to 10°

nodal lines) need less energy for rotors having a multiple of 3 poles, i.e. a total of 6, 12, 18 etc. poles, and so on.

Figure 4.9 displays sets of results for different pole shapes. Narrow poles (10°), have higher frequencies of vibrations, wider poles (20°) – lower frequencies. Generally, the shape of the pole itself has less significant influence on overall results, they just follow the pattern. Again, curves depicting frequency modes are much more influenced by the shape of rotor.

Finally, 9 pairs of rotors were modeled in order to check the influence of thickness on natural vibration frequencies (Fig. 4.10 and 4.11). Results indicate that thickness increase results in equivalent shift of eigenfrequencies, i.e. twofold increase of thickness yields twofold increase of frequencies of all vibration modes.

Modal analysis resulted in frequencies above 100 kHz, while rotation frequency of micromotors is in order of kilohertz. Moreover, frequencies rise rapidly with increase of thickness of a rotor. This concludes that under normal operating conditions the rotor is far from resonance conditions and the motor can be operated successfully.

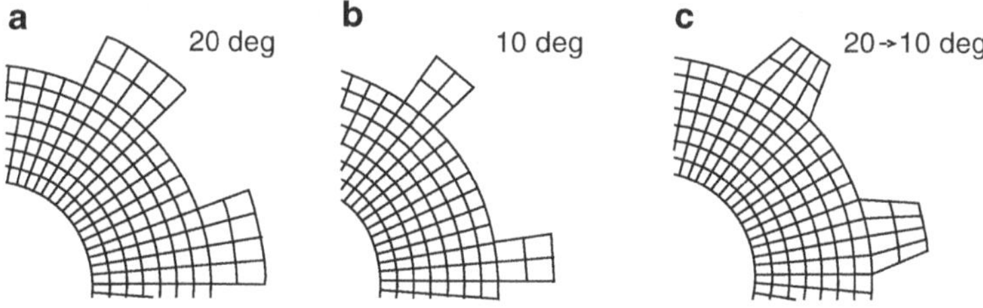

Fig. 4.9 Different shapes of poles modeled

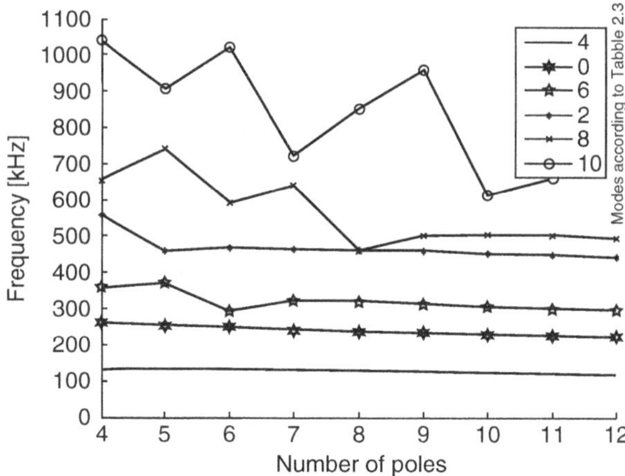

Fig. 4.10 Vibration modes of 2 μm thickness rotor

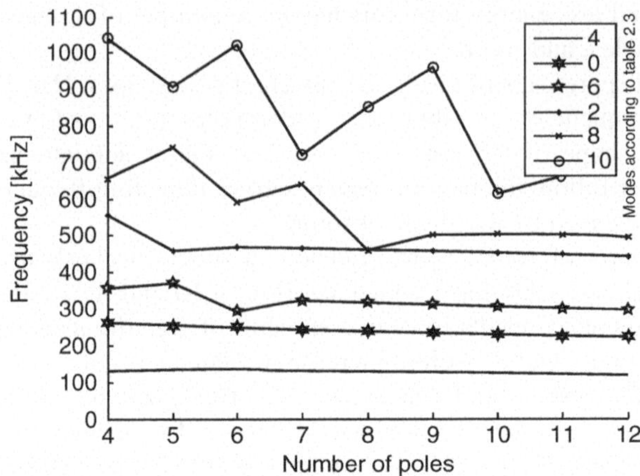

Fig. 4.11 Vibration modes of 4 μm thickness rotor

4.4 Dynamics of MEMS Structures in Viscous Medium

At micro-scale, unlike at macro-scale, damping by surrounding medium cannot be neglected and has great influence on dynamics of MEMS components. There are important both friction losses between molecules in the moving media (viscosity), and between the media and the solid surfaces (squeeze film and slide film damping). Sometimes these losses are so huge, that they prevent functioning of a motor entirely. In order to minimize them, quite often in laboratory conditions the micro-actuators are run in vacuum [8–10] Practically this is not always feasible, making it necessary to develop new micromotors and corresponding models that are valid at any pressure of surrounding gas.

The major damping influencing the micromotor is due to the internal friction of the flowing gas, namely viscosity, in small clearances between the moving elements. At low pressures or in ultra-thin films, the gas rarefaction effect and the molecular interaction with the surfaces effectively change the viscosity. This is the case in MEMS structures, where the gap height is only a few micrometres and the gas pressure is much smaller than atmospheric pressure. Thus, an accurate gas-flow model is important in the device design.

The damping of the structural material is usually orders of magnitude lower than the viscous damping and is generally neglected. The resulting damping in the dynamical system is dominated by the internal friction of the air between the rotor and the substrate and poles. These viscous damping effects can be captured by using two general damping models: Couette flow damping and squeeze film damping.

Couette flow damping occurs when two plates of an area A, separated by a distance y_0, slide parallel to each other (Fig. 4.12). Assuming a Newtonian gas, the Couette flow damping coefficient can be approximated as

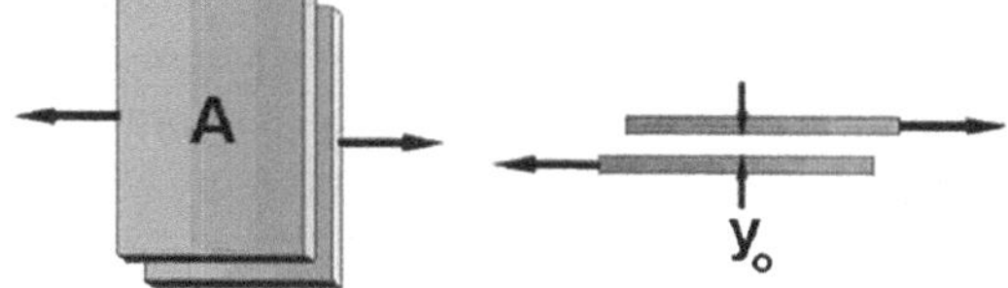

Fig. 4.12 Illustration of Couette flow damping between two sliding plates

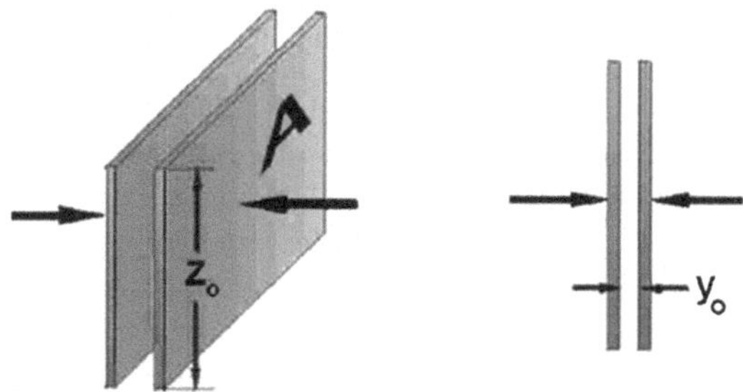

Fig. 4.13 Illustration of squeeze-film damping between two plates

$$c_{Couette} = \mu_p p \frac{A}{y}, \tag{4.3}$$

where $\mu_p = 3.710^{-4}$ kg/m^{2}s·torr is the viscosity constant for air, p is the air pressure, A is the overlap area of the plates, y is the plate separation.

Squeeze film damping occurs when two parallel plates approach each other and squeeze the fluid film in between (Fig. 4.13). Squeeze film damping effects are more complicated, and can exhibit both damping and stiffness effects depending on the compressibility of the fluid. Using the Hagen-Poiseuille law, squeeze film damping can be modeled as

$$c_{Squeeze} = \mu_p p \frac{7Az_o^2}{y^3}. \tag{4.4}$$

Utilizing the presented damping models, the total damping can be expressed as the sum of damping due to Couette flow between the rotor and the substrate, the damping due to Couette flow between the rotor and stator poles and the damping due to Couette flow between the rotor and stator axle and the squeeze film damping arising with vertical rotor vibrations.

$$c_x = \mu_p p \frac{A_{rotor}}{y} + \mu_p p \frac{Z_s A_{pole}}{y_{gap}} + \mu_p p \frac{A_{rotoraxis}}{y_{gap2}} + \mu_p p \frac{7A_{rotor} z_o^2}{y^3} \tag{4.5}$$

More accurate slide-film damping models can be generated considering the gas rarefaction effects at low pressures and narrow gaps, kinetic gas models, or plate motions that propagate into the fluid with rapidly diminishing steady-state amplitude. Including the non-linear effects of squeeze-film damping together with computational fluid dynamics simulations will also improve the accuracy of the damping model.

Air damping varies with the pressure level from high vacuum to atmosphere and can be divided into three states [11]:

1. Intrinsic damping region: in this region, the air damping is negligible.
2. Molecular region: the damping is caused by independent collision of noninteracting air molecules with the moving surface.
3. Viscous region: air acts as a viscous fluid.

Viscous drag torque on the micromotor can be modeled using the boundary element method (BEM). In this method, the Stokes equations are recast into an integral form relating the velocities and tractions on the boundary of the fluid. For known surface velocities, the integral equations are solved numerically for the surface tractions by discretising the boundary into a series of elements and solving the resulting linear system of equations. This method requires only the boundary of the fluid to be discretised rather than the entire fluid domain, thereby greatly reducing the size of the problem relative to finite element method.

In a moving fluid, the Reynolds number is a measure of the ratio of the viscous forces to the inertia forces, so that a low Reynolds number implies that the fluid inertia is negligible. For a rotating disc, the Reynolds number is given by:

$$\mathrm{Re}_1 = \rho R^2 \frac{\omega}{\mu}, \tag{4.6}$$

where ρ, μ and R are the fluid density and viscosity and the disc's radius and angular velocity respectively. Using the values for these given for air $\rho = 1.2\ \mathrm{kg/m^3}$, $\mu = 1.83 \times 10^{-5}$, $R = 150\ \mu\mathrm{m}$ and $\omega = 630$ rad/s (100 turns/s) gives a Reynolds number of the order of 1. For a disc rotating above a fixed plate a Reynolds number should be used

$$\mathrm{Re}_2 = \rho R h \frac{\omega}{\mu}, \tag{4.7}$$

where h is the elevation of the disc.

Using this expression (with $h = 2\ \mu\mathrm{m}$), the Reynolds number for this micromotor is reduced to 0.1. The analysis of Phan-Thien indicates that the first-order correction terms for the fluid inertia in the rotating disc flow are of 0 (Re_2^2), and are therefore very small for the present problem. This justifies modeling of the flow by the Stokes equations. The start up torque, however, could be significantly different from the Stokes value, depending on the manner in which the motor was started up. This can only be modeled correctly if the inertial term $\rho \partial u/\partial t$ is included in the equations of motion [12].

The viscous drag of micromotor can be measured experimentally using parameter estimation from stroboscopic photographs. Numerical calculations of a viscous drag can be carried out using a two-dimensional (axisymmetric) finite element analysis with the micromotor represented by a disc, and using a three-dimensional finite element analysis to model the micromotor as a finned propeller.

4.5 Modeling of Micromotor in Viscous Media

Air in narrow gaps between components of microdevices can not be neglected because at the micro scale, the air in these gaps acts as a fluid that damps the movement of mechanical parts.

Micromotor modeling in surrounding air is, however, much more complicated as it comprises at least two individual bodies that have an ever changing contact point. At microscale no ball bearings are possible, thus bushings have to be used. Moreover, bushings create too much friction, thus a reasonable clearance between a shaft and a ring is necessary. This clearance also depends on possibilities of lithographic equipment and is usually at least 0.5 μm and up to 5 μm. Five micrometer clearance comprises close to 10% of radius of a 70 μm axle. This is a very big distance that in usually inconceivable in macroscopic world and results in difficultly predictable motion of the rotor.

Thus, depending on motion, the rotor can touch stator axle at a point (Fig. 4.14) or even a line (Fig. 4.15), or can be entirely not in contact at some time instant.

Modeling of all motor features in 3-D electrostatic field, including a surrounding viscous medium will not give easy practically ascertainable results as there are too many physical effects involved that are way too complicated to model at the same time.

Thus, first step of analysis of motor in viscous medium is done by modeling of a rotor freely suspended in air above ground plate without gravity. In these conditions, the rotor does not move, but can only vibrate, so the squeeze-film analysis is performed. The eigen-frequency analysis of the rotor in viscous medium was done using the specialized FE software COMSOL Multiphysics (Fig. 4.16). This model of the rotor couples the squeezed-film gas damping, which is modeled with the linearized Reynolds equation, together with displacements in the rotor at its eigenmodes.

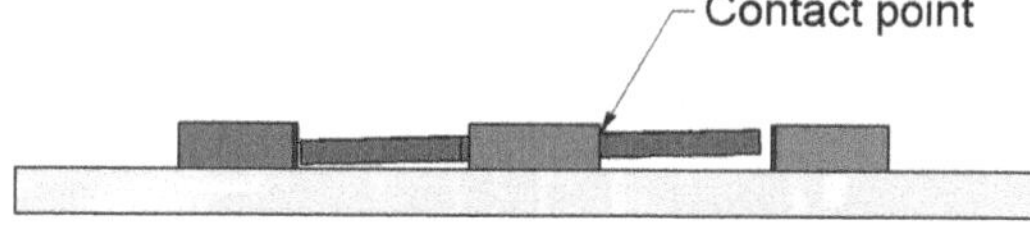

Fig. 4.14 Rotor and axle touching at a point

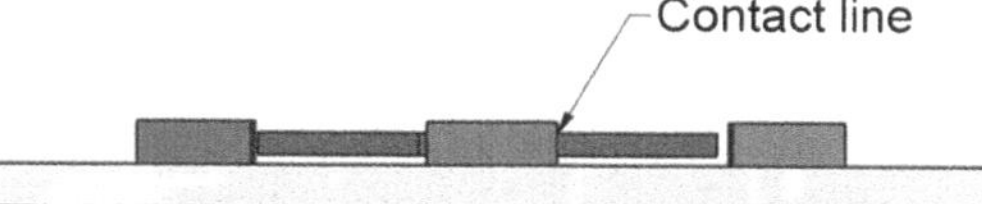

Fig. 4.15 Rotor and axle touching along a line

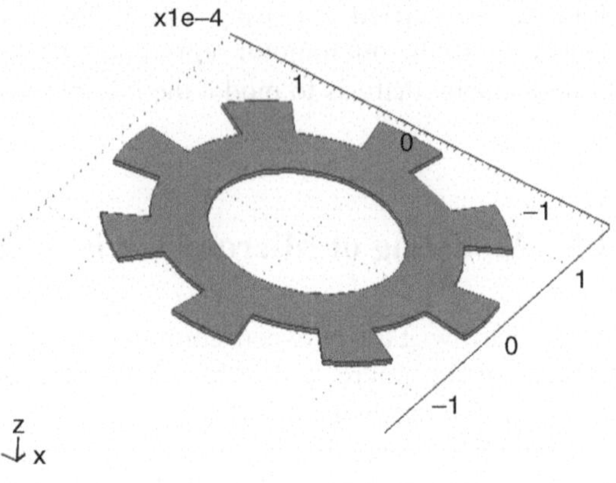

Fig. 4.16 3-D view of micro-rotor

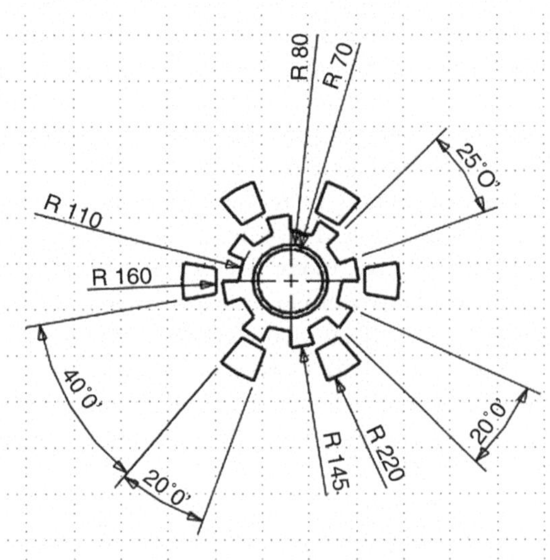

Fig. 4.17 Micro-motor dimensions

Dimensions of the modeled rotor were taken according to manufactured samples (Fig. 4.17). The model includes a narrow gap between the rotor and the substrate (Fig. 4.18).

First COMSOL model analysis was done by setting the thickness of the gas layer to 5 μm, rotor thickness to 4 μm and the gas pressure set to atmospheric pressure of 1 bar. The model was solved and the first few normal modes and eigenfrequencies were observed and analyzed. Figure 4.19 and 4.20 show the deformed shape and the pressure distribution on the surface of the rotor at the first and second eigenmodes respectively.

Further, the pressure of gas was varied and changes in the frequencies of modes were studied. Figure 4.21 represents relationship between the pressure in the gap and

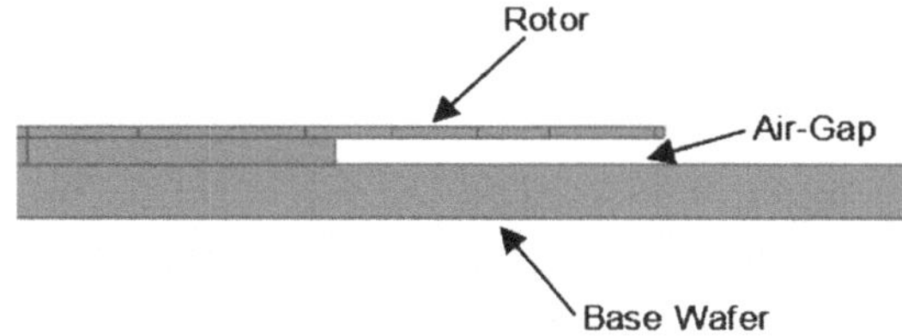

Fig. 4.18 Gap between rotor and wafer

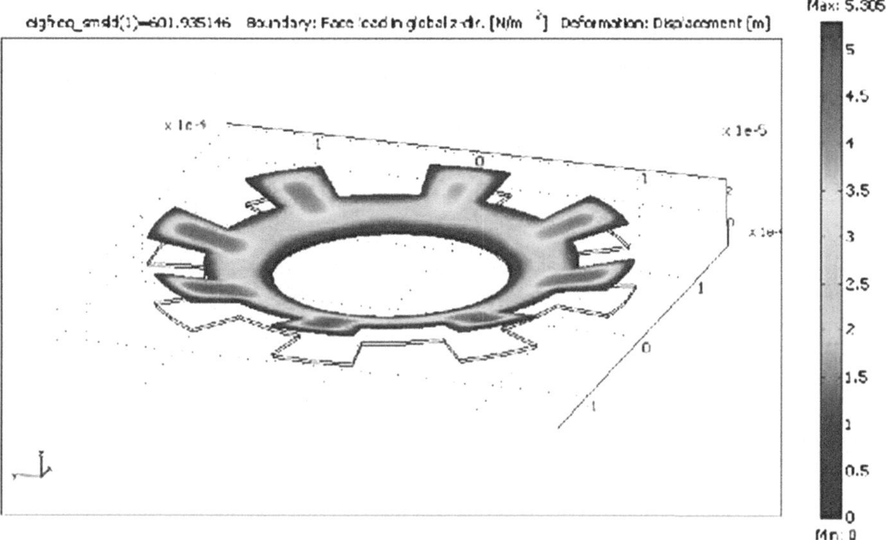

Fig. 4.19 Pressure distribution on the surface of rotor at first eigenmode

the resulting mode frequencies at the specified gap thickness. Apparently, pressure and frequency have an inversely proportional relationship: by increasing gas pressure, frequencies at each mode tend to decrease. This proves the fact that the air surrounding the micro-rotor influences the dynamic response of the rotor and is actually acting as damping mechanism. As seen from the graph the influence is not linear.

In order to verify that the model was correctly prepared and at vacuum the thickness of the gap has no effect on the eigenfrequencies since there is no gas acting as a damping fluid, the pressure was set to zero (vacuum) and the thickness of the gap was varied. This setup had to result in no changes in the frequencies because there was no air in the gap to affect the rotor. This relation between the gap height (at vacuum) and the frequencies of eigenmodes is shown in Fig. 4.22. It is obvious from the graph that the model generates equal values of frequencies at different gap thicknesses, and therefore, indicates that the model was properly designed.

It is worth mentioning that the eigenfrequencies at the first three modes obtained from this research coincide with the eigenfrequencies obtained in a previous research in which an eigenfrequency analysis of the rotor was performed in vacuum without considering the gap between the rotor and the base ground.

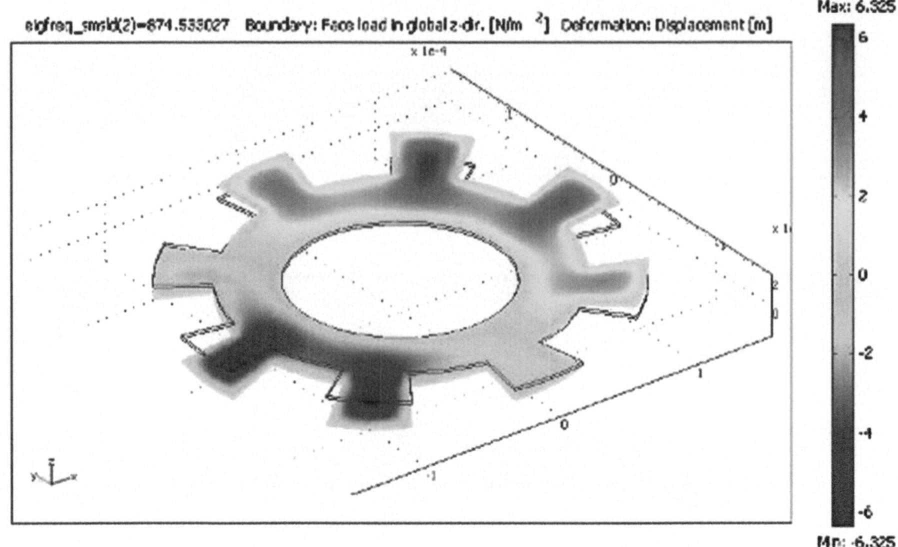

Fig. 4.20 Pressure distribution on the surface of rotor at second eigenmode

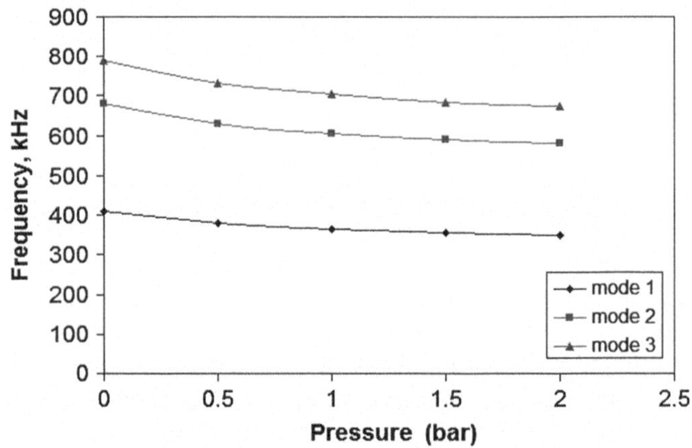

Fig. 4.21 Dependence of rotor natural frequencies on the pressure in the gap

4.6 Control of a Micromotor

There is plenty of micromotor designs created and many micromotors produced, and all of them have different number of poles, different geometry and dimensions. Obviously, there cannot be one that performs best, but thorough analyses were seldom reported. Using analytical and FE methods, it is possible to assess the electrostatic torque of a variable capacitance electrostatic micromotor. In this section, moment of

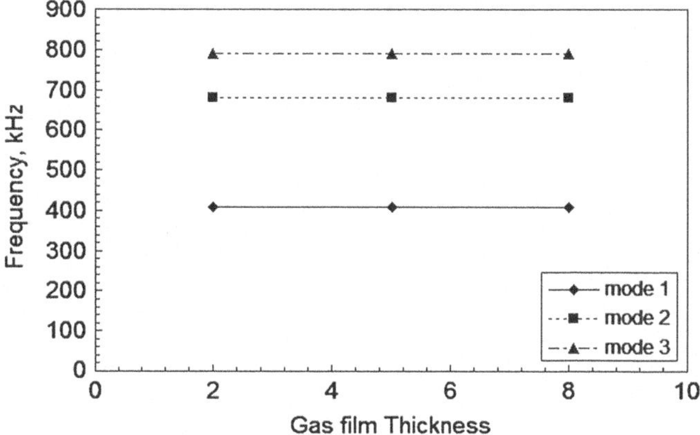

Fig. 4.22 Proof of viscous media model in vacuum

rotation produced by electrostatic forces is calculated as a function of a single pole rotor position, by varying a number of selected geometrical relations.

From the 2-D numerical simulation, it is shown that the geometric parameters have a major influence on the moment of rotation produced by a single pole of an electrostatic micromotor and thus care should be taken when designing a micromotor. But even high moment of rotation produced by a single pole is not enough to create a high torque micromotor. Depending of the motor configuration different switching sequences and techniques need to be applied. To solve these problems, further modeling is necessary based on results of analysis of moment of rotation produced by a single pole.

4.6.1 Analytical Model of a Micromotor

A simplified analytical model is a good tool for fundamental calculations. Different constructions can be simulated readily without highly time-consuming FE modeling and simulations. Parameters in the analytical model are changed quickly, but the disadvantage is that motor poles are reduced to point charges.

A motor is composed of arbitrary number of stator and rotor poles. When there is a potential difference between any poles, an electrostatic force is created that acts on rotor. Only tangential force acting on rotor pole will create moment of rotation. Normal force will only pull rotor towards stator, creating harmful friction between rotor and its axle.

The rotor is grounded and the stator electrodes are charged ''on'' and ''off'' in a sequence. For example, considering electrostatic micromotor shown in Fig. 4.23, suppose stator q_1 is turned ''on'' while stators q_2 and q_3 are ''off''. The attracting force between charge q_1 and rotor Q will cause rotation of the rotor clockwise.

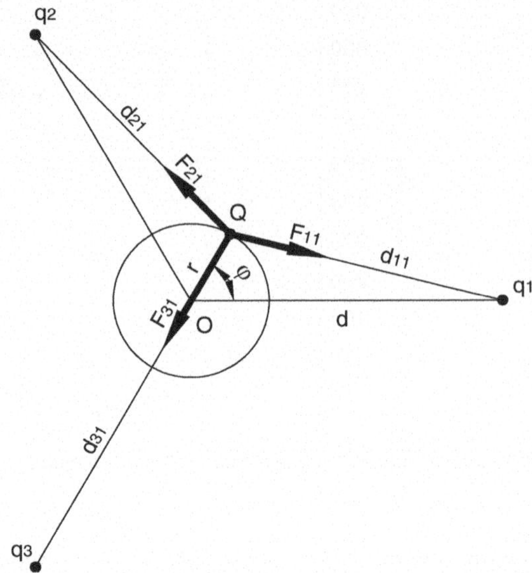

Fig. 4.23 Free-body diagram of a simplified 3-stator 1-rotor pole micromotor

Then, q_1 is turned ''off'' and q_3 is turned ''on'' causing Q to be attracted by q_3. Similarly, before the rotor makes a complete cycle, q_3 is turned ''off'' and q_2 is turned ''on'' causing attraction forces between Q and q_2. This is carried out over and over again that creates turning of the rotor.

In order to create dynamical model of the motor, firstly forces created between poles need to be analyzed. Stator poles will further be regarded as $I = 1, \ldots, m$ and rotor poles $j = 1, \ldots, n$. Force between first stator and first rotor poles is F_{11}, second stator and first rotor – F_{21} and so on. Similarly distance between poles is d_{11}, d_{21}, etc.

Generally, force between any stator and rotor pole is

$$\vec{F}_{ij} = k_c \cdot \frac{Q \cdot q_i}{d_{ij}^2}, \tag{4.8}$$

assuming that stator and rotor poles are point charges, where Q – is the charge of rotor, q_i – charge of i-th stator, and $k_c = 8.9875 \times 10^9$ Nm^2/C^2 is Coulomb's constant.

Both distance d_{ij} and its tangential component d_{ijt} depend on angle between rotor and stator $\angle\varphi_{ij}$. According to cosine law,

$$d_{ij}^2 = r^2 + d^2 - 2rd\cos\phi_{ij}. \tag{4.9}$$

As depicted in Fig. 4.24, solving for $\angle\varphi$ sine:

$$\sin\phi = \frac{d_{ijt}}{d}, \tag{4.10}$$

$$\sin\mu = \frac{d_{ijt}}{d_{ij}} = \frac{d\sin\phi}{d_{ij}}, \tag{4.11}$$

and expressions for ∠φ cosine:

$$\cos\phi = \frac{r + d_{ijn}}{d}, \tag{4.12}$$

$$\cos\mu = \frac{d\cos\phi - r}{d_{ij}}, \tag{4.13}$$

where angle $\angle\mu$ is necessary to find ratio between force and its tangential and normal components:

$$\sin\mu = F_{ijt}/F_{ij}, \tag{4.14}$$

$$\cos\mu = F_{ijn}/F_{ij}, \tag{4.15}$$

$$F_{ijt} = F_{ij}\sin\mu = F_{ij}\frac{d\sin(-\phi)}{d_{ij}}, \tag{4.16}$$

$$F_{ijn} = F_{ij}\cos\mu = F_{ij}\frac{d\cos\phi - r}{d_{ij}}. \tag{4.17}$$

According to common standards, moment of rotation is negative if it acts in counter-clockwise direction, thus the angle $\angle\varphi$ in Eq. 4.16 is considered to be negative.

So far expressions containing ∠φ were considered and valid only for angle between first stator and rotor poles (Fig. 4.24). In order to extend this analysis to arbitrary number of stator and rotor poles, the following variables need to be added: β – angle between stator poles, γ – angle between rotor poles (Fig. 4.25) that are expressed as:

$$\beta = \frac{2\pi}{m}, \tag{4.18}$$

$$\gamma = \frac{2\pi}{n}. \tag{4.19}$$

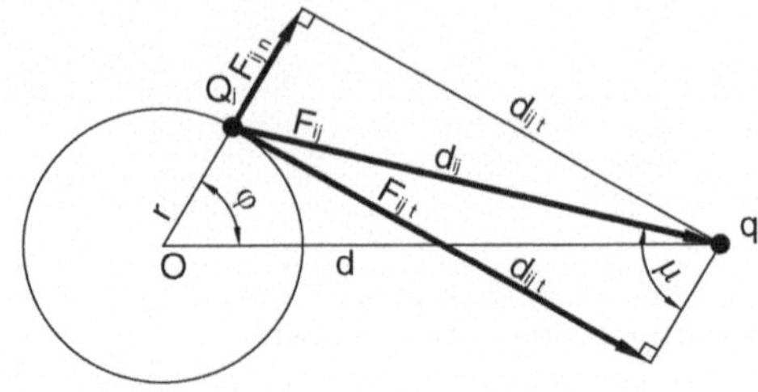

Fig. 4.24 Tangential and normal forces created between a pair of stator and rotor poles

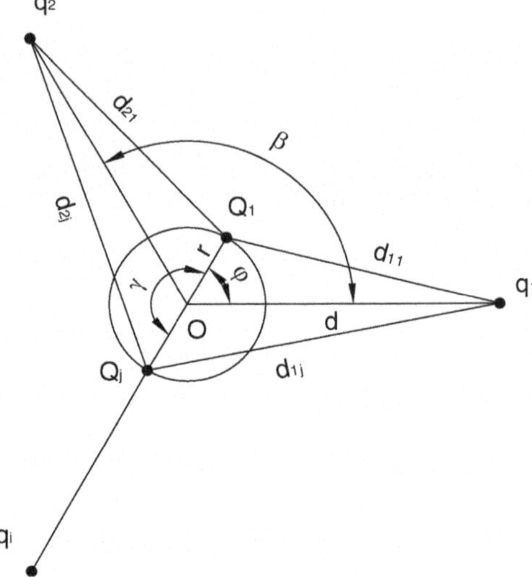

Fig. 4.25 Micromotor diagram describing arbitrary number of stator and rotor poles

Solving the trigonometry problem gives a matrix of angles between corresponding rotor and stator poles $\{\varphi\}$ which are expressed as:

$$\begin{pmatrix} \phi_{11}^2 & \phi_{12}^2 & \phi_{23}^2 & \cdots & \phi_{1n}^2 \\ \phi_{21}^2 & \phi_{22}^2 & \phi_{23}^2 & \cdots & \phi_{2n}^2 \\ \phi_{31}^2 & \phi_{32}^2 & \phi_{33}^2 & \cdots & \phi_{3n}^2 \\ \cdots & \cdots & \cdots & \phi_{ij}^2 & \cdots \\ \phi_{m1}^2 & \phi_{m2}^2 & \phi_{m3}^2 & \cdots & \phi_{mn}^2 \end{pmatrix} =$$

$$= \begin{pmatrix} -\phi & -\gamma-\phi & -2\gamma-\phi & \cdots & -(n-1)\gamma-\phi \\ \beta-\phi & \beta-\gamma-\phi & \beta-2\gamma-\phi & \cdots & \beta-(n-1)\gamma-\phi \\ 2\beta-\phi & 2\beta-\gamma-\phi & 2\beta-2\gamma-\phi & \cdots & 2\beta-(n-1)\gamma-\phi \\ \cdots & \cdots & \cdots & (i-1)\beta-(j-1)\gamma-\phi & \cdots \\ (m-1)\beta-\phi & (m-1)\beta-\gamma-\phi & (m-1)\beta-2\gamma-\phi & \cdots & (m-1)\beta-(n-1)\gamma-\phi \end{pmatrix} \tag{4.20}$$

Combining (4.9) and (4.20) a distance between two poles is expressed as:

$$d_{ij} = \sqrt{r^2 + d^2 - 2rd\cos((i-1)\beta - (j-1)\gamma - \phi)}. \tag{4.21}$$

According to the Eqs. 4.16, 4.17 and 4.20, tangential and normal components of electrostatic force between i-th stator and j-th rotor pole are:

$$F_{ijt} = k_c \frac{Q \cdot q_i}{d_{ij}^2} \cdot \frac{d \sin \phi_{ij}}{d_{ij}} = k_c \frac{Qq_i d \sin \phi_{ij}}{d_{ij}^3}, \tag{4.22}$$

$$F_{ijn} = k_c \frac{Q \cdot q_i}{d_{ij}^2} \cdot \frac{d \cos \phi_{ij} - r}{d_{ij}} = k_c \frac{Qq_i \left(d \cos \phi_{ij} - r \right)}{d_{ij}^3}. \tag{4.23}$$

A simplified 3-stator 1-rotor pole motor, like shown in Fig. 4.23, has pole distances as depicted in Fig. 4.26 that are calculated according to (4.21). Calculated tangential forces according to the analytical model (4.22) are shown in Fig. 4.27. The bold curve denotes total developed torque of the micromotor. As seen from the graphs, distances between poles change similarly to sine law as the motor rotates, and forces produce peeks when they are close to corresponding stator poles. After passing the angle of pole alignment, tangential forces change their sign. At this moment charge should be removed from a stator pole, otherwise negative torque would be created that slows a rotor down (Fig. 4.28).

Naturally, depending on pole switching sequence, a rotor pole can be attracted by some stator poles, thus total force acting is:

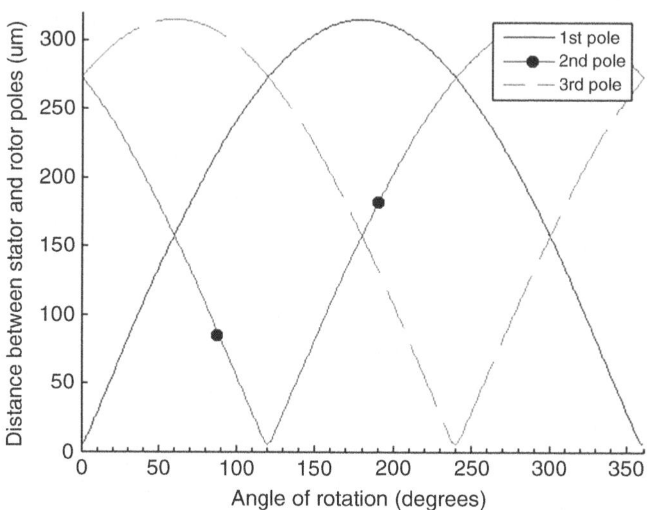

Fig. 4.26 Distance between sets of poles of a simplified 1-rotor, 3-stator pole micromotor

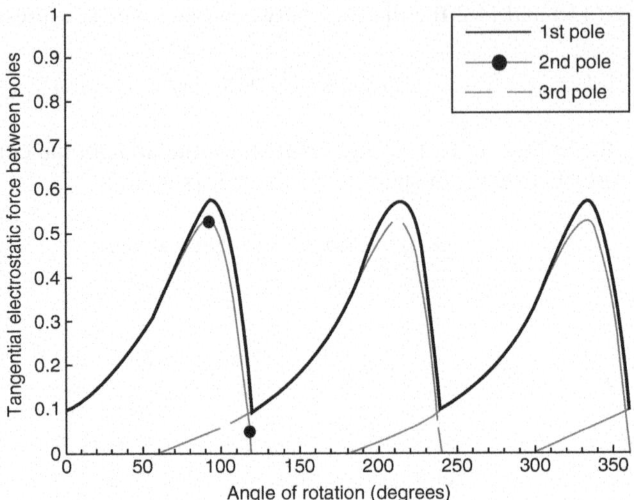

Fig. 4.27 Moment of rotation produced by poles and total torque of the motor

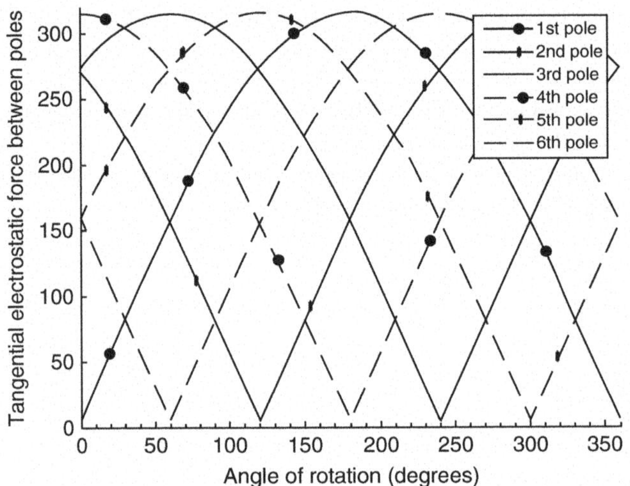

Fig. 4.28 Distance between sets of poles of a simplified 2-rotor, 3-stator pole micromotor

$$\vec{F_j} = \sum_{i=1}^{n} \vec{F_i}. \tag{4.24}$$

Torque produced by a single pole:

$$M_{oi} = r \sum_{j=1}^{m} F_{ijt}. \tag{4.25}$$

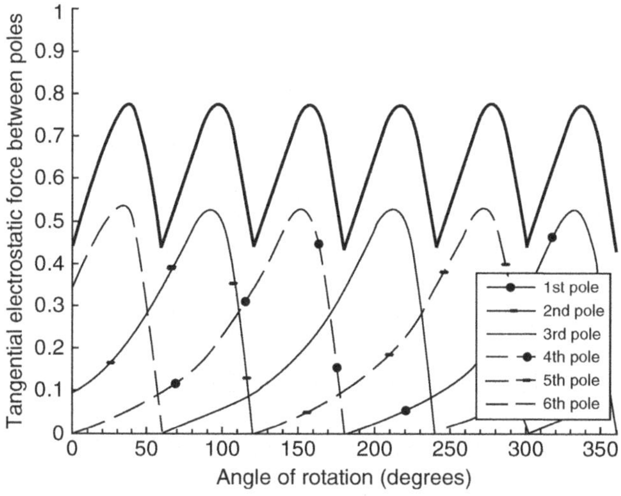

Fig. 4.29 Moment of rotation produced by poles and total torque of the motor

Torque produced by the motor:

$$M_o = r \sum_{i=1}^{n} \sum_{j=1}^{m} F_{ijt}. \tag{4.26}$$

A similar motor, but having two rotor poles, produces a significantly higher torque (Fig. 4.29). This confirms results received in finite element modeling. Naturally, torque ripple becomes smaller, too.

Dynamic behavior of micromotor can be characterized by the following equation:

$$I\ddot{\phi} + c\dot{\phi} = M_0(\phi, q_i, t), \tag{4.27}$$

where I – the moment of inertia of the disk, c – the damping coefficient and M_0 – the moment of the disk about center point that depends on charge of stator poles and angle of rotation at a time moment.

The moment of inertia of a disk is:

$$I = \frac{m \cdot r^2}{2}. \tag{4.28}$$

Combining (4.26–4.28), the equation describing dynamics of micromotor is:

$$\frac{m \cdot r^2}{2}\ddot{\phi} + c\dot{\phi} = r \sum_{i=1}^{n} \sum_{j=1}^{m} \left(k_c \frac{Q \cdot q_i(t) \cdot d \sin \phi_{ij}}{d_{ij}^3} \right). \tag{4.29}$$

Using this differential equation it is possible to predict how a motor will perform at different sequences of pole switching, also its spin-up and slow down processes.

4.6.2 Simplified Electrostatic–Mechanical Scheme

First, an interaction between two oppositely charged bodies (micromotor poles) had to be analyzed. A 2-D Cartesian space was used, where one pole is immovable (stator pole) and the other one (rotor pole) can move along an axis without energy losses.

If the poles were oppositely charged point bodies, the movable body would tend to get as close as possible to the unmovable body (Fig. 4.30). Similar results happen when two comparable size poles interact: their equilibrium point is reached when centers of both poles align (Fig. 4.31). Normal force F_y does not create motion, whereas tangential force F_x is responsible for lateral movement.

Though it might seem confusing, but increase in size of poles does not increase tangential force. This is proved by splitting bodies into finite elements. Each element attracts opposite element with force that depends on the square distance between them, according to Coulomb's law. Thus, the bigger is the distance, the smaller is the force. Elements A1 and A2 are attracted to all B elements with some specific tangential forces (Fig. 4.32). But tangential forces between A3, A4, A5 and all B elements cancel out (Fig. 4.33). Though normal force increases between bigger poles, tangential force remains nearly the same. Thus, tangential force can be increased only by increasing potential difference between poles or by decreasing

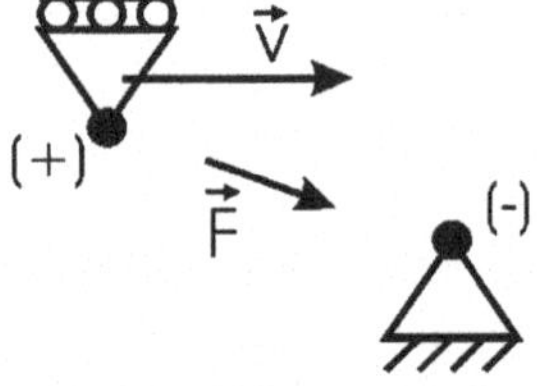

Fig. 4.30 A force created between two oppositely charged point masses

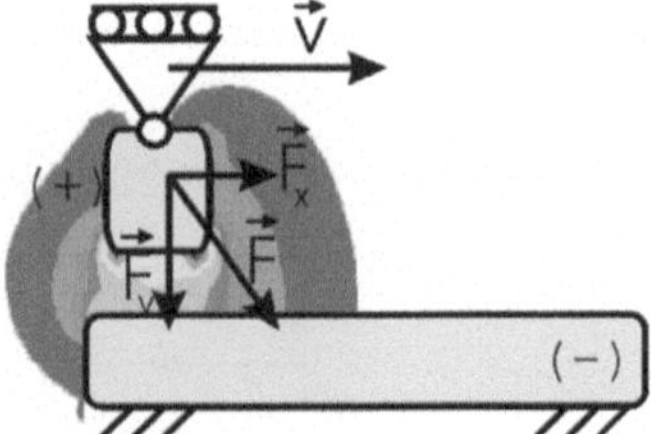

Fig. 4.31 A simplified electrostatic–mechanic scheme of two oppositely charged poles

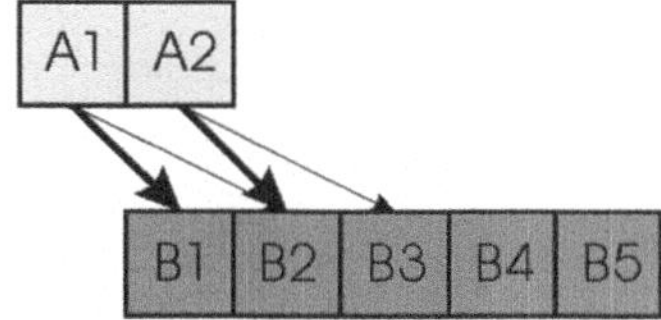

Fig. 4.32 FE structure of poles of different sizes. B4 and B5 segments are attracted with only negligible force as compared to B1 and B2

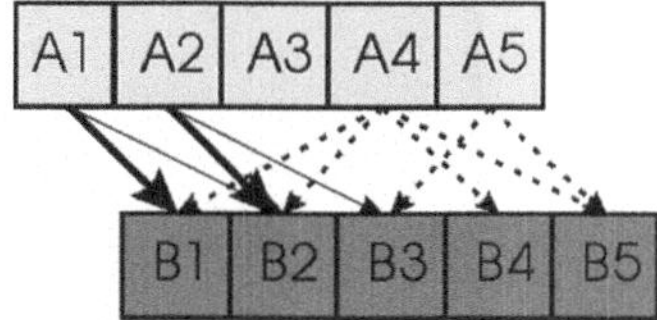

Fig. 4.33 Pull force between equal size poles. Segments A1, A2 create most tangential force, while the other ones cancel out or act oppositely

the distance between them. In a 3-D space it can also be increased by increasing the area of interacting poles.

Different micromotor designs are possible depending on required accuracy, speed of rotation, power, etc. A specific switching sequence can be calculated for any motor. Theoretically, a motor can work using any number of stator and rotor poles, though practically the most efficient solution is to interconnect stator poles to have three phases. Two phases are not enough to determine the direction of rotation. Three phases determine rotation direction accurately like in most popular asynchronous motors in macro-world. Four phases are not necessary because they make control electronics and motor design complicated without any obvious advantages. Thus obviously the number of stator poles needs to be a multiple of three: 3, 6, 9, 12, etc.

In order to analyze resulting interactions between stator and rotor poles, a complete 2-D ANSYS model of the system was created, where potential difference, distances and geometry of rotor and stator were easily varied. The program simulates developed electrostatic field at a specified angular step between the rotor and an active stator pole and outputs quantitative value of produced torque (Figs. 4.34 and 4.35).

4.6.3 Moment of Rotation Produced by a Single Pole

The torque of a motor is not steady and varies significantly as poles interact. If a task is to keep the torque at maximum throughout whole 360° turn cycle, a specific switching sequence needs to be developed for a motor. But in order to quantitatively evaluate the torque created by the motor, a moment of rotation produced by a single pole needs to be analyzed first.

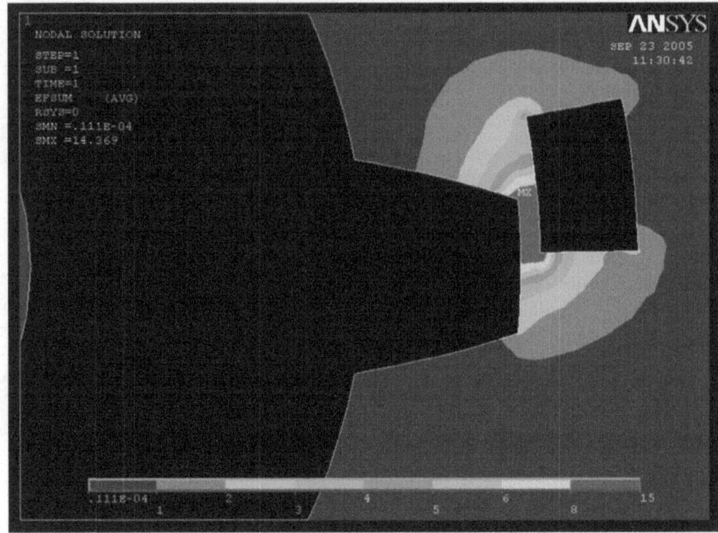

Fig. 4.34 FE modeling of electrostatic field between rotor and stator poles

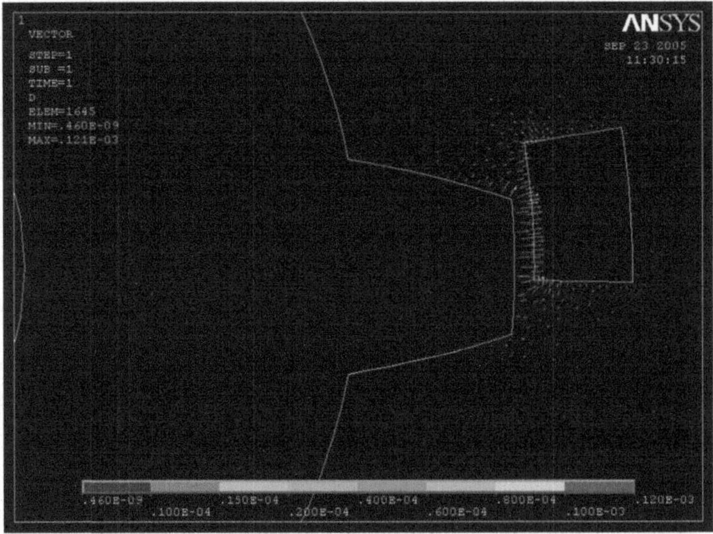

Fig. 4.35 FE vector modeling of electrostatic field between rotor and stator poles

When centers of stator and rotor poles are aligned, no moment of rotation is produced. In further analysis this position is assumed to be of a zero degree angular difference. As the angular difference is increased, the moment of rotation increases and then suddenly drops down. The drop is a result of an interaction of equipotential electrostatic fields of two adjacent rotor poles when stator is a middle between them (Fig. 4.36).

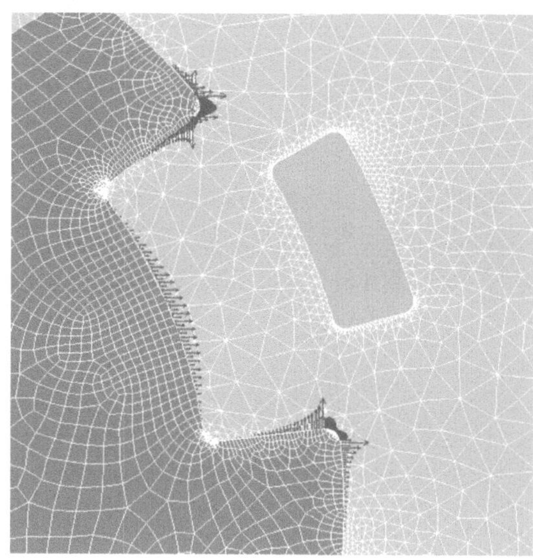

Fig. 4.36 Electrostatic forces developed when stator pole is in a middle of two rotor poles

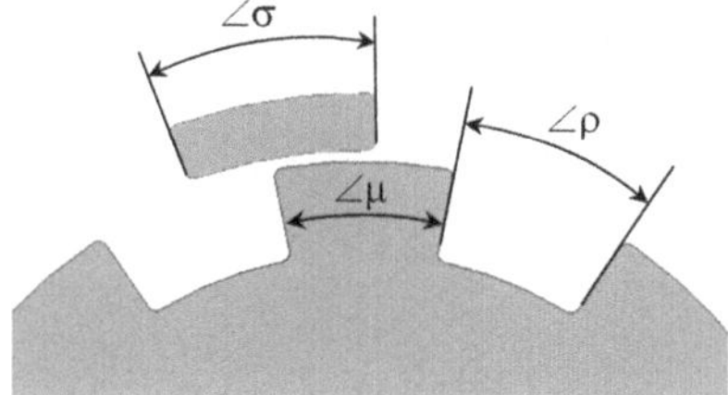

Fig. 4.37 Dimensions used in the analysis

A full pole cycle starts at zero tangential force when poles are aligned and ends also at zero tangential force in a middle between two rotor poles. Thus, the cycle of a single pole can be always expressed as 360° divided by a number of rotor poles. The angle taken by a rotor pole will further be designated by $\angle\mu$, angle of the rotor pole gap – $\angle\rho$ and angle of the stator pole – $\angle\sigma$ (Fig. 4.37). All stator and rotor radii are kept constant in further analysis if not specified expressly.

Torque curve depends on the ratio between angles of rotor pole $\angle\mu$ and rotor pole gap $\angle\rho$. In order to get maximum torque throughout the cycle, the integral of the curve needs to be as big as possible.

To analyze this, the number of poles was kept constant and the size of a rotor pole was varied. FE analysis showed that the highest moment of rotation integral curve was created if $\angle\mu/\angle\rho \approx 1.0$, i.e. when the rotor pole and the gap between poles fill the same angle.

Having this ratio less or more than a unity creates only a part of maximum curve. There is a small torque difference for ratios 0.9–1.1, but outside of this interval, the torque drops significantly (Fig. 4.38).

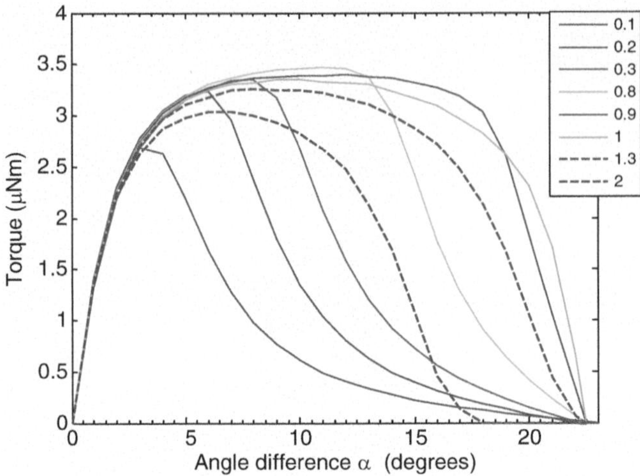

Fig. 4.38 Torque depending on the ratios between rotor pole and rotor pole gaps

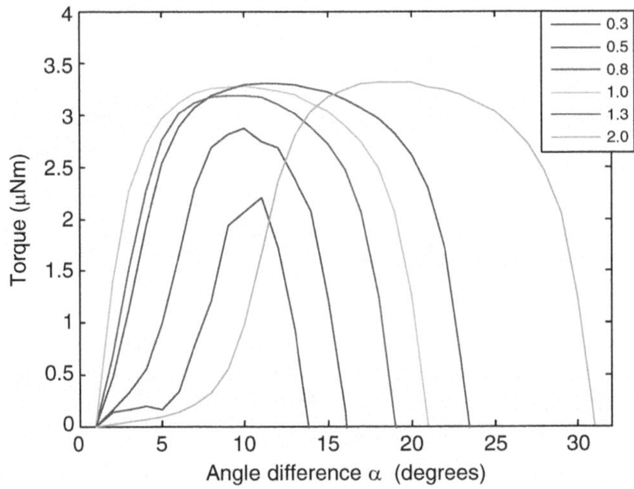

Fig. 4.39 Torque created by a rotor pole of a defined ratio between stator and rotor pole sizes

As concluded above, the best results were achieved if the angle of a pole gap was nearly equal to stator pole, thus $\angle\mu$ is set equal to $\angle\rho$. In further analysis, rotor size was varied and the pole cycle was considered no longer constant, so only the average torque created by a pole was significant for this study. Best case situation would be pole cycle as small as possible and the torque as high as possible. Once again, the best results were achieved when actually all three angles were equal (Fig. 4.39). It is important to note that the last curve (ratio 2.0) is of biggest integral value, but the average value is lower than the others.

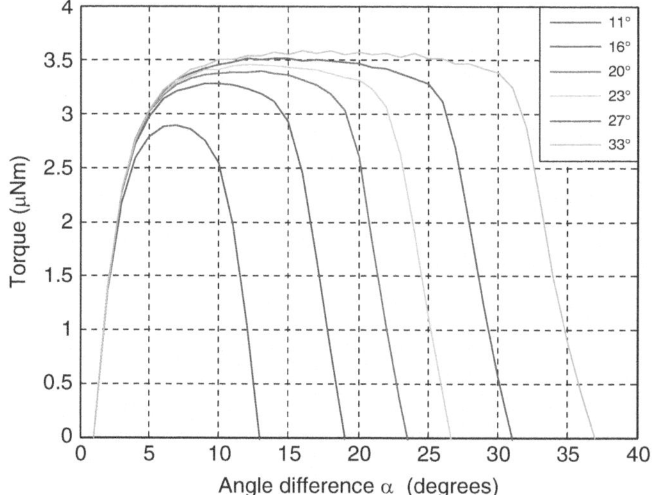

Fig. 4.40 Torque created by a single rotor pole of a specified angular size

Thus, the final pole geometry conclusion is that all specified dimensions have to be equal to get the maximum torque $\angle\mu = \angle\rho = \angle\sigma$.

Depending on the number of poles a rotor has, the torque produced by a pole can vary significantly (Fig. 4.40). The average torque of a single pole is directly proportional to the size of the pole. Confirming conclusions established in earlier paragraph, pole size has a small influence on maximum value produced, which is only slightly higher for bigger poles.

Finally, FE analysis was carried out to determine the influence of stator-rotor gap (Fig. 4.41). It is inversely proportional, thus by decreasing the gap twice the torque will increase twice (Fig. 4.42).

As seen from modeling results above, serious care must be taken in designing even slightest details of a micromotor such as its pole form. Sizes of stator and rotor poles should be the same in order to achieve higher moment of rotation. Gaps between interacting poles should be as small as possible.

4.6.4 Modeling of Torque of a Micromotor

A MATLAB program was created to evaluate performance of motors having different geometries. The most important input parameters are moment of rotation produced by a single pair of poles and geometrical specifications of the motor itself. The requested speed of rotation must be entered, too. When all necessary data is inputted, the program displays micromotor configuration (Fig. 4.43a) and a corresponding torque–angle curve produced by a pair of poles (Fig. 4.43e). It is important to note that motor configurations can have different moment of rotation

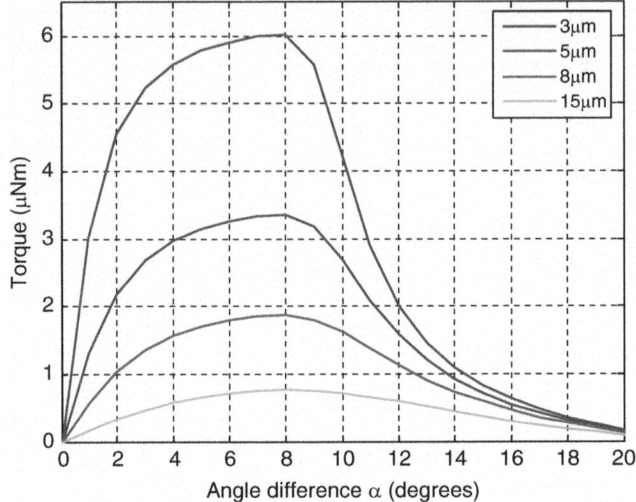

Fig. 4.41 Torque created by a motor having specified gap between rotor and stator

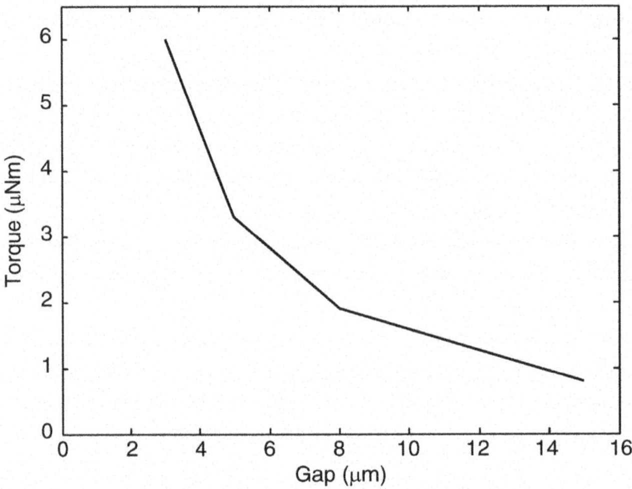

Fig. 4.42 Dependence between torque and stator-rotor gap

curves produced by a pair of poles. For this reason the program selects a corresponding curve from its database of previously simulated ones, though generally if the size of poles is unaltered, same curve can be also applied for analysis of a micromotor having different number of poles.

The program outputs an animation sequence of how stator poles are energized by tagging poles in brighter color, required number of phases, period lengths, duty cycles, minimal angular difference graph and finally, momentum and average motor torques.

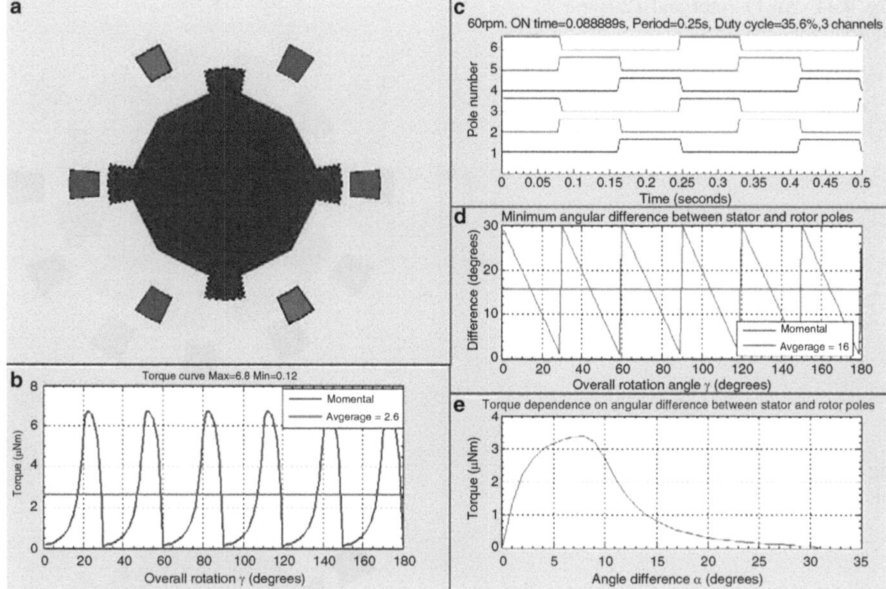

Fig. 4.43 A MATLAB program for analysis of torque of a micromotor

Depending on the number of stator poles they are automatically combined into phases. By default, three phases are always preferred, i.e. three, six or nine poles make three phases, four poles – four phases, five poles – five phases and so on. The advantages of having three phases are obvious: IC electronics is much simpler and the rotor gets a balanced loading. The time diagram of phase switching is showed in a graph (Fig. 4.43c). The program calculates which poles have to be switched "ON" and how long in order to create maximum torque for the requested speed of rotation. Thus the following data is outputted: period and duration of switching duty cycle, duration of "ON" state and the required number of channels (phases).

There is also a complementary graph that shows the smallest angular distance between a pair of poles (Fig. 4.43d). This is important because very small torque is produced if poles are too far away from each other. As seen in Fig.4.43e nearly no torque can be produced for angular differences above 15° or even less. This causes uneven torque. Worst case scenario can happen if the motor is stopped when a distance between poles is large and possibly it will not start because of high friction or stiction forces.

The produced torque curve and an average torque value over a complete 360° rotation is shown in Fig. 4.43b.

Different modeling cases are displayed in Figs. 4.44 and 4.45. The program calculates distances from stator poles to adjacent rotor poles and simulates "ON" condition only if the created moment of rotation is in the required direction. Both figures display groups of adjacent poles that cause rotation in counter-clockwise direction. Pole switching diagrams for the illustrated micromotors are presented in

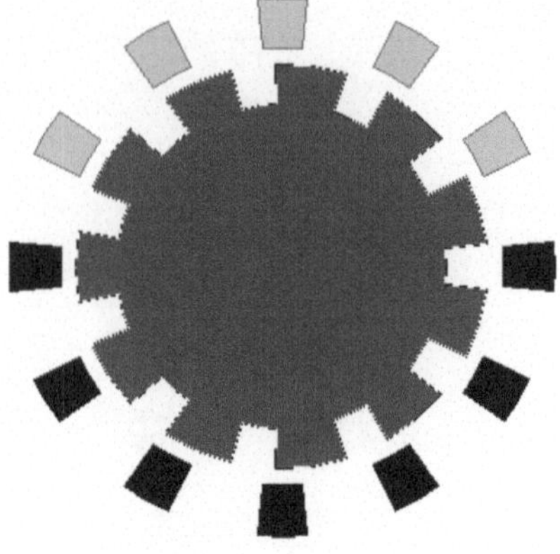

Fig. 4.44 An 11-rotor and 12-stator micromotor scheme

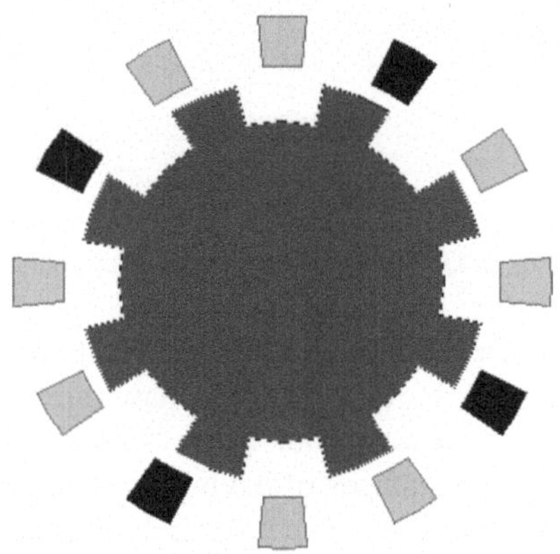

Fig. 4.45 An 8-rotor and 12-stator micromotor scheme

Figs. 4.46 and 4.47. The first motor would need to have 12 phases, whereas the second motor would need only three phases. Modeling shows that the second option is a more efficient solution, though the first motor has higher accuracy and could be used as a stepper motor.

Using the program a number of micromotor constructions were quantitatively analyzed from 3 to 16 stator and rotor pole combinations. They were evaluated according to average torque and number of phases necessary for control. Only motors capable to work with three phases were selected.

As seen from 3-stator pole (Figs. 4.48–4.50) and 6-stator pole (Figs. 4.51–4.53) motors, increasing the number of rotor poles has a small negative effect on the overall torque as maximum and minimum torque values drop. This is typical for all motor configurations. As the total number of poles increase, the ripple becomes smaller, because total switching is more often and smaller peaks are created.

Increasing the number of stator poles results in significantly higher torque and smaller ripple (Figs. 4.54 and 4.55).

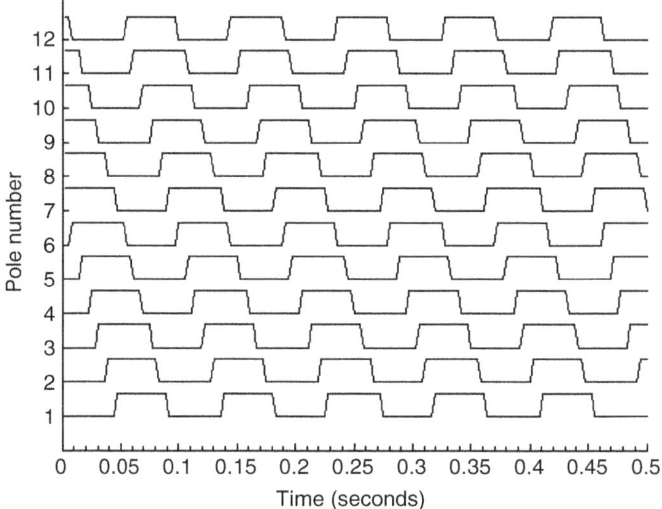

Fig. 4.46 Pole switching diagram for 11-rotor and 12-stator micromotor

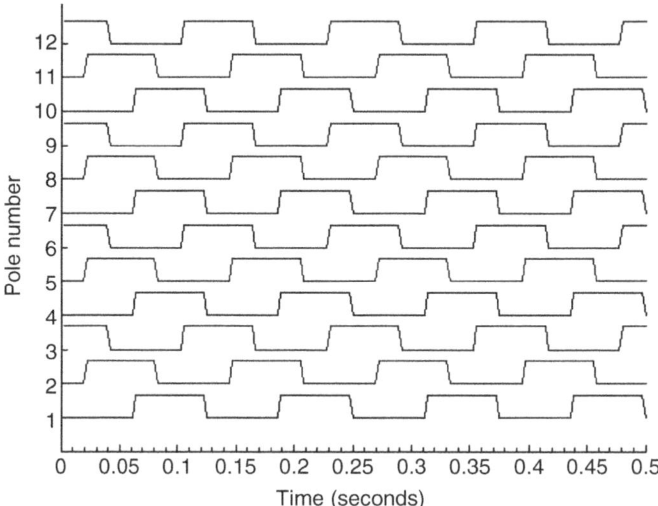

Fig. 4.47 Pole switching diagram for 8-rotor and 12-stator micromotor

Same effects were consistently repeated for other micromotor configurations having different number of poles (Figs. 4.56–4.61). This data was summarized and presented in Figs. 4.62 and 4.63 the following conclusions were drawn:

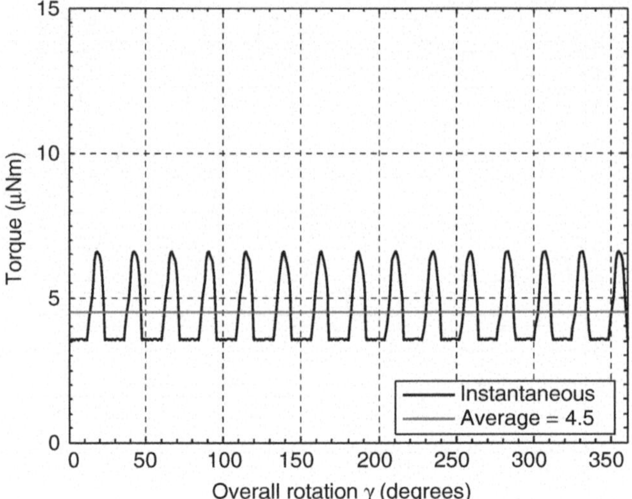

Fig. 4.48 Torque created by 3-stator, 5-rotor pole motor

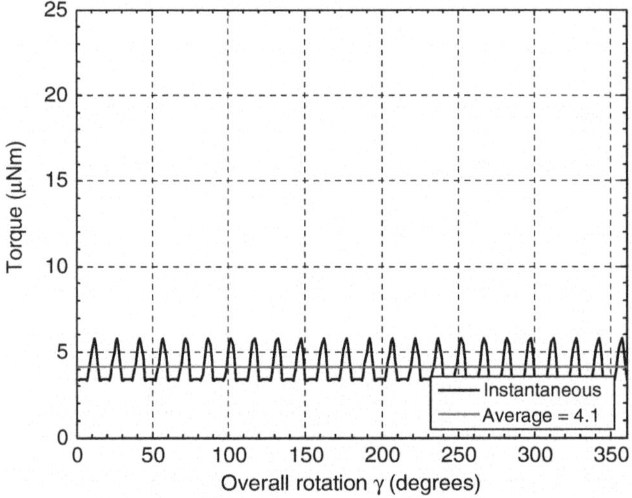

Fig. 4.49 Torque created by 3-stator, 8-rotor pole motor

1. By decreasing the number of rotor poles, the torque slightly increases. For instance, if the number of rotor poles is increased 2.5 times, the average torque decreases by around 15%. Exclusion of this regularity is that the number of rotor poles should not be less than 4.
2. A change in number of stator poles gives a proportional change in torque. For instance, by increasing the number of poles twice of a three-phase motor, the torque will increase twice, too.

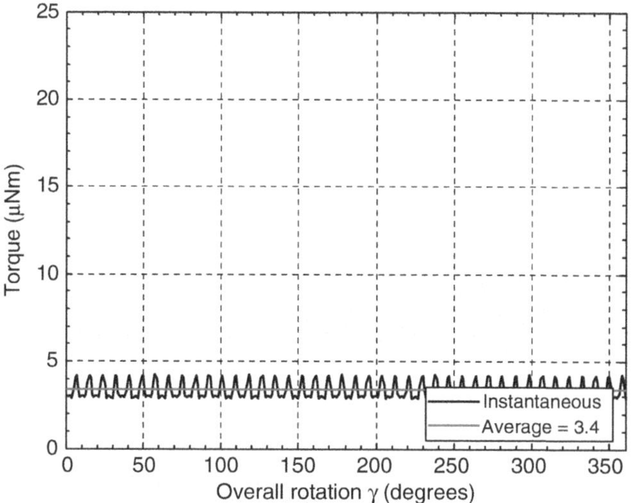

Fig. 4.50 Torque created by 3-stator, 14-rotor pole motor

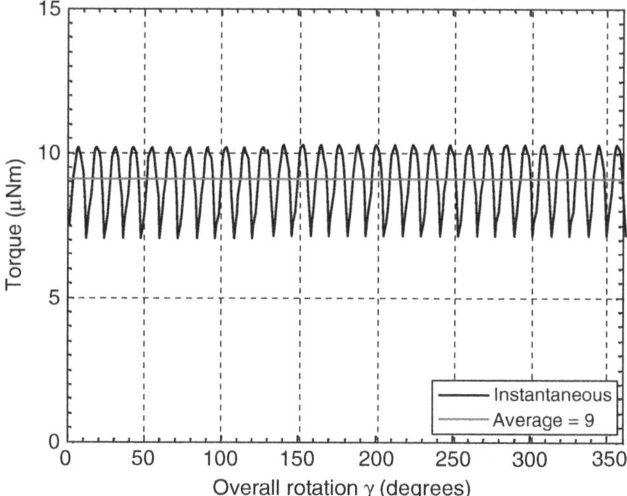

Fig. 4.51 Torque created by 6-stator, 5-rotor pole motor

Thus, irrelevantly of number of phases, the biggest torque is created at maximum number of stator poles and minimum number or rotor poles.

4.6.5 *Micromotor Torque and Switching Sequences*

An electrostatic micromotor exhibits pulsating torque as seen from previous graphs. The torque has average and ripple components that are more or less distinct for any

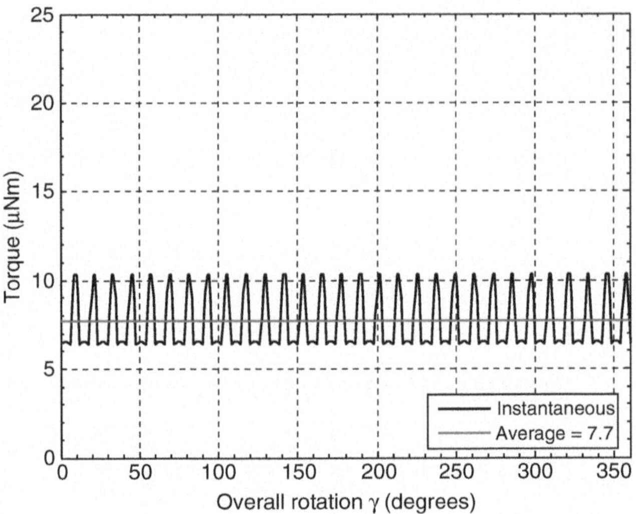

Fig. 4.52 Torque created by 6-stator, 10-rotor pole motor

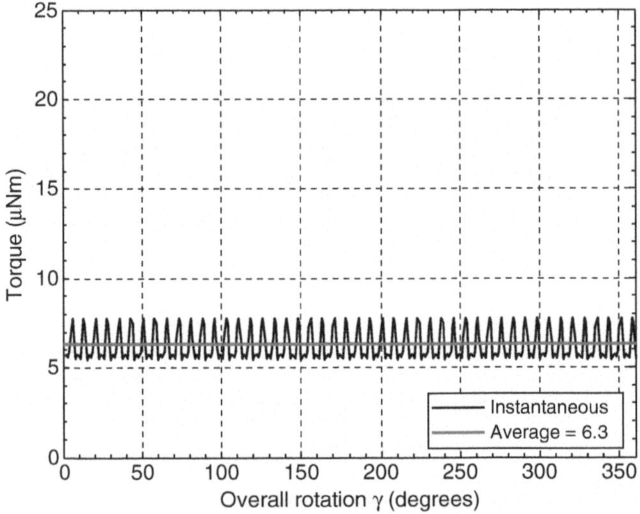

Fig. 4.53 Torque created by 6-stator, 16-rotor pole motor

kind of micromotor construction (Fig. 4.48). The average torque, T_{av}, is found using common mathematical averaging methods of the generated torque over 360°. The torque ripple T_{rpl} is defined as

$$T_{rpl} = \frac{T_{\max} - T_{\min}}{T_{av}}. \tag{4.30}$$

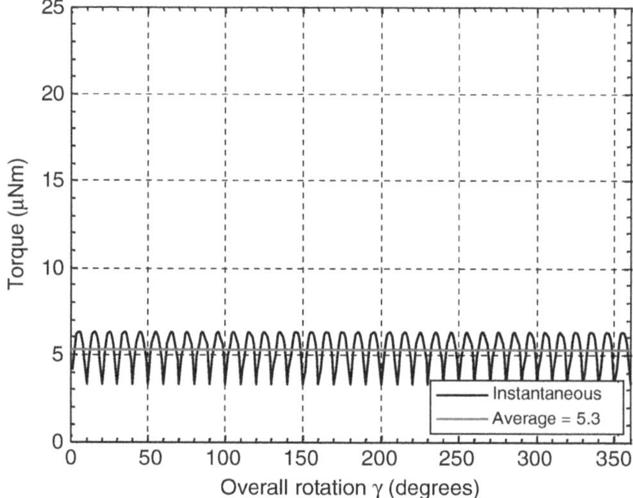

Fig. 4.54 Torque created by 4-stator, 9-rotor pole motor

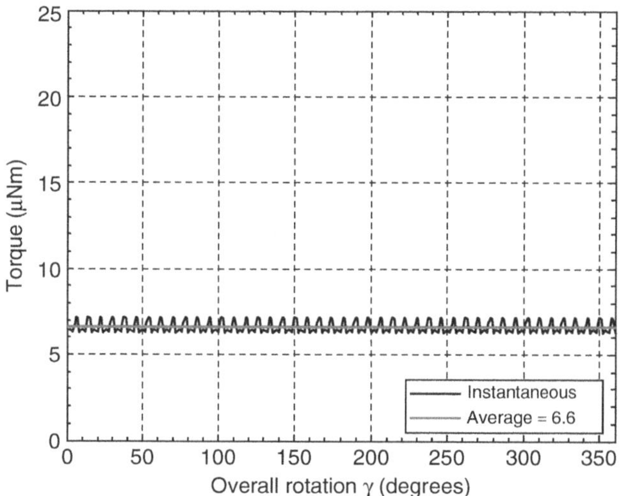

Fig. 4.55 Torque created by 5-stator, 9-rotor pole motor

Usually the most important feature of a micromotor is its average torque. Torque ripple is an inevitable effect of such motor that can cause rotor vibrations and should be kept low. There are some studies regarding it, for instance, an article by V. Behjat [6] states that the optimal micromotor producing the lowest torque ripple is found by setting relative stator pole width to 65% and relative rotor pole width to 40%. These results might be correct, though as seen here from the results of models

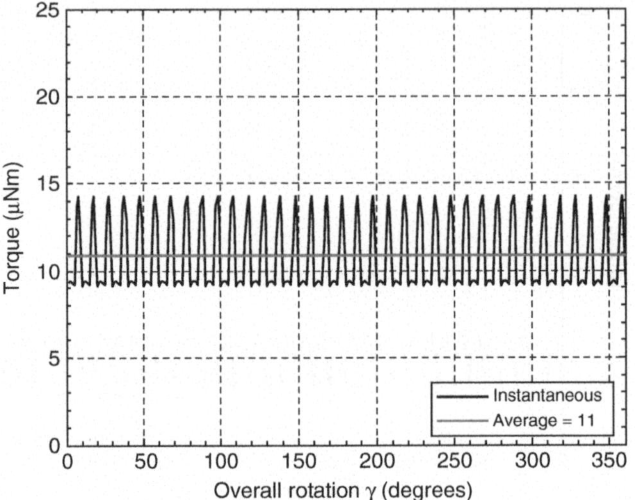

Fig. 4.56 Torque created by 9-stator, 12-rotor pole motor

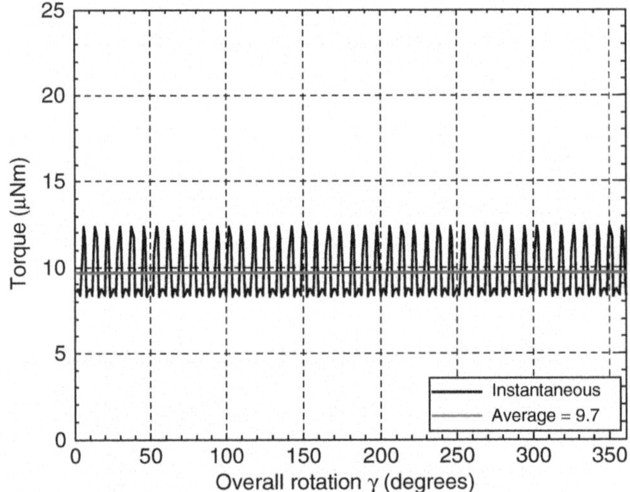

Fig. 4.57 Torque created by 9-stator, 15-rotor pole motor

above, ripple can generally be decreased by increasing the total number of poles (Table 4.4). This increases also the average torque. Unfortunately, higher number of poles results in more complicated structure of a micromotor, but pole sizes should be kept according to previous recommendations in order to keep the torque to the maximum. Micromotors that have smaller number of poles and higher ripple should be separately analyzed for possible resonant frequencies of self-excited vibrations.

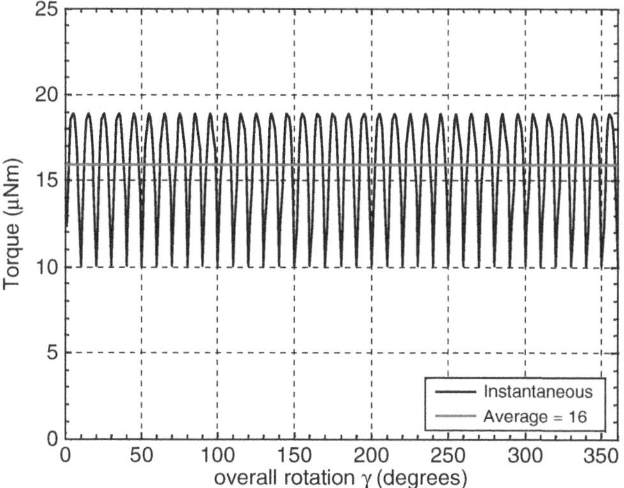

Fig. 4.58 Torque created by 12-stator, 9-rotor pole motor

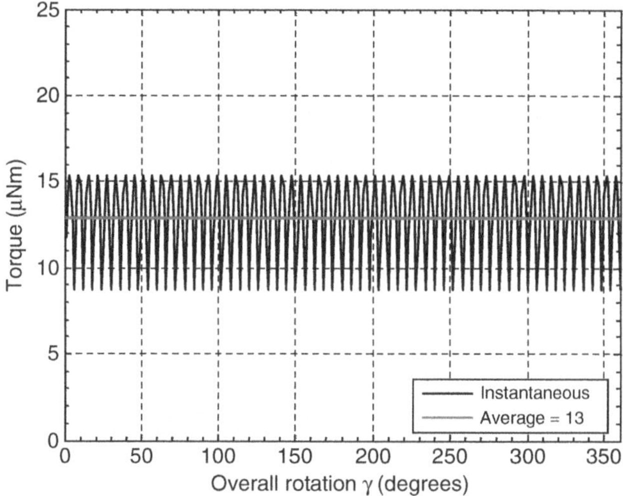

Fig. 4.59 Torque created by 12-stator, 15-rotor pole motor

Depending on configurations of motor, different switching sequences should be used to spin a rotor in counter clockwise direction. The simplest three pole motors are controlled by energizing poles in clockwise direction sequence (1-2-3- . . . first, second, third pole). Poles of more complicated motors 6-pole motors, are energized in pairs: first and fourth pole at the same time, followed by second and fifth pole, etc). Similarly nine pole motors are energized by three poles at a time: 1,5,9; 2,6,10;

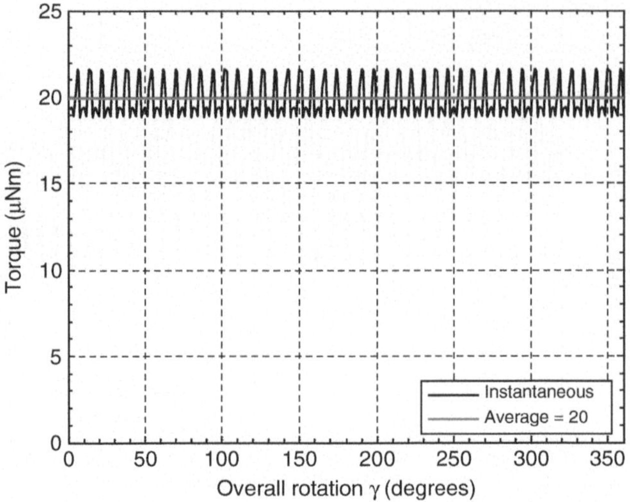

Fig. 4.60 Torque created by 15-stator, 9-rotor pole motor

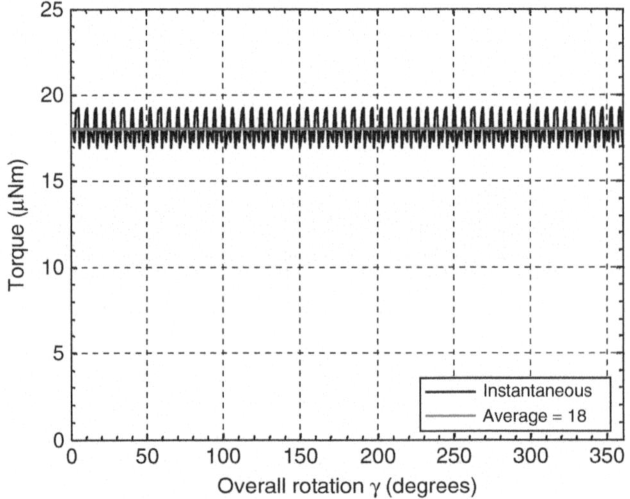

Fig. 4.61 Torque created by 15-stator, 12-rotor pole motor

etc. Depending on configuration, the switching should be performed in clockwise or counter clockwise direction (Table 4.5).

If there is a difference of one pole between numbers of stator and rotor poles, for example six stator and five rotor poles, then the switching sequence is always sequential: 1, 2, 3, 4, etc.

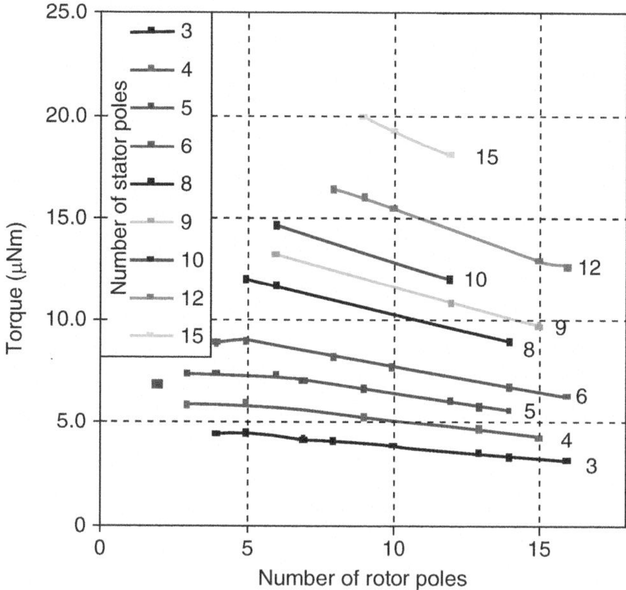

Fig. 4.62 Torque dependence on the number of rotor and stator poles

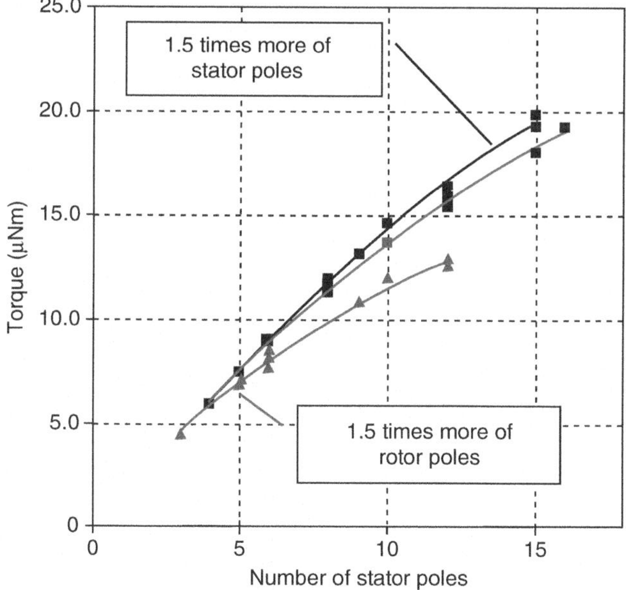

Fig. 4.63 Torque dependence on the number of stator poles

Table 4.4 Influence of features of a micromotor on its ripple

Stator poles	Rotor poles	Average torque	Max torque	Min torque	Ripple
3	5	4.5	6.6	3.5	0.69
3	8	4.1	5.8	3.3	0.61
3	14	3.4	4.3	2.9	0.41
6	5	9	10	7	0.33
6	10	7.7	10	6.4	0.47
6	16	6.3	7.8	5.4	0.38
9	12	11	14	9.1	0.45
9	15	9.7	12	8.3	0.38
12	9	16	19	10	0.56
12	15	13	15	8.7	0.48
15	9	20	22	19	0.15
15	12	18	19	17	0.11

Table 4.5 Control sequences of various micromotor configurations

Number of stator poles	Number of rotor poles	Sequence of pole energizing
3	5	1-2-3 – 1-2-3- ...
3	8	1-2-3 – 1-2-3- ...
3	14	1-2-3 – 1-2-3- ...
6	5	1-2-3-4-5-6 – 1-2-3-4-5-6- ...
6	10	1,4-2,5-3,6 – 1,4-2,5-3,6- ...
6	16	1,4-2,5-3,6 – 1,4-2,5-3,6- ...
9	12	3,6,9-2,5,8-1,4,7 – 3,6,9-2,5,8-1,4,7- ...
9	15	1,4,7-2,5,8-3,6,9 – 1,4,7-2,5,8-3,6,9- ...
12	9	1,5,9-2,6,10-3,7,11-4,8,12 – 1,5,9-2,6,10-3,7,11-4,8,12- ...
12	15	4,8,12-3,7,11-2,6,10-1,5,9 – 4,8,12-3,7,11-2,6,10-1,5,9- ...
15	9	1,6,11-2,7,12-3,8,13-4,9,14-5,10,15 - 1,6,11-2,7,12-3,8, 13-4,9,14-5,10,15- ...
15	12	5,10,15-4,9,14-3,8,13-2,7,12-1,6,11 - 5,10,15-4,9, 14-3,8,13-2,7,12-1,6,11 - ...

Using the given switching sequences motor poles are grouped into three phases and controlled by external programmable device, according to Fig. 4.64.

Based on the results of previous FE analysis, the micromotor can be designed according to one of the following requirements.

Simplest construction. This kind of micromotor would have a small number of poles, small torque, high torque ripple, simple construction and simple electronics. This motor would have three or six rather big stator poles connected into three phases (Fig. 4.65). Three stator poles would give a radial unbalanced load, which leads to even higher torque ripple, more vibrations and wear. Anything more than one pole per phase produces balanced load, making a 6-pole construction a better choice. The construction is relatively simple, though electronic circuit requires

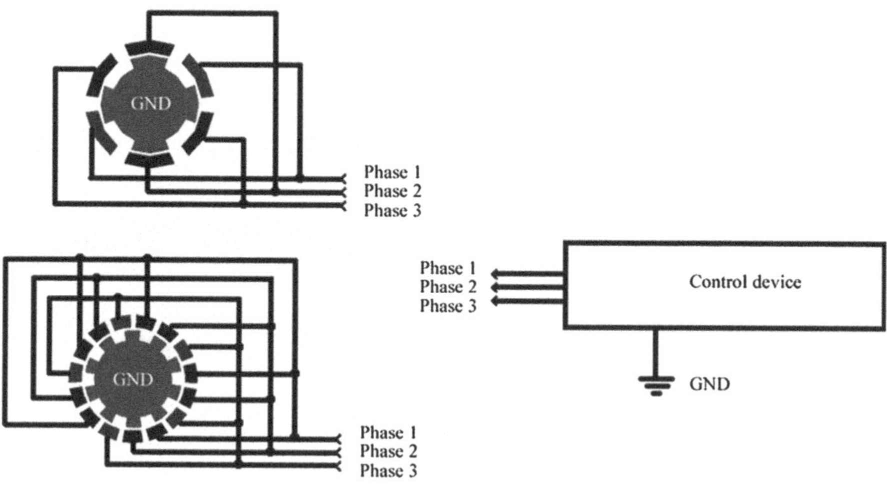

Fig. 4.64 Different micromotor connection to a control device

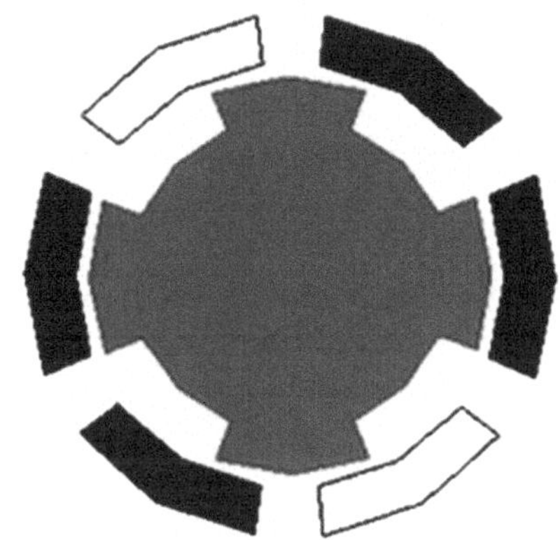

Fig. 4.65 Simplest construction motor: 3 phases, 6 poles. Two opposite poles are energized simultaneously

minimum two layers. The main drawback is that this kind of micromotor produces small torque.

Highest torque. This kind of micromotor would have a large number of poles, highest torque, low ripple, complicated construction and usually same simple electronics. The load is radially balanced in all cases. Generally such a motor would have more than two rotor poles per phase (Fig. 4.66).

Highest accuracy. This kind of micromotor can be treated as a stepper motor, because one pulse on a phase creates smallest angle of rotation. The control electronics of such motor is complicated because depending on accuracy it requires a rather big number of phases. Simple and effective switching sequences for such motors are if numbers of rotor and stator poles differ by one (Fig. 4.67). If run at a constant speed, the torque ripple of this motor is low.

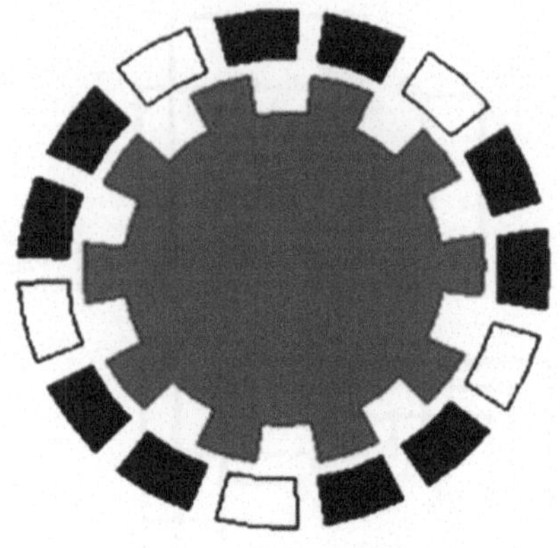

Fig. 4.66 Highest torque motor: 3 phases, 15 poles. Five poles are energized simultaneously

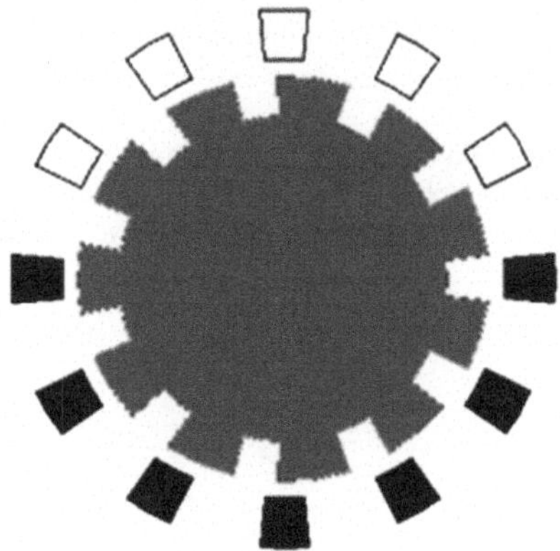

Fig. 4.67 Highest accuracy motor: 12 phases, 11 poles. Five poles are energized simultaneously

The micromotor constructions mentioned above might have one possible drawback: only some rotor poles are subjected to electrostatic force from stator poles at the same time instant. Alternatively, a motor can be designed which has all rotor poles subjected to attraction force constantly (Fig. 4.68). This design is not according to conclusions drawn earlier, rotor and stator poles are not of equal angular size and the moment of rotation created by a single pole is considerably smaller.

Simulation shows that a decrease in ratio between rotor and stator pole angular sizes gives a nearly proportional decrease in created torque (Fig. 4.69). Taking this into account, every pole creates smaller attraction force but the number of active poles is higher. A working procedure of such a micromotor would be similar to one of a constantly running electromagnetic stepper motor, because an active stator pole is always a little ahead of following rotor pole. Also, similarly to the stepper motor, the position of the rotor can be easily and accurately controlled. Unfortunately, the number of layers is higher and control electronics of such micromotor is much more

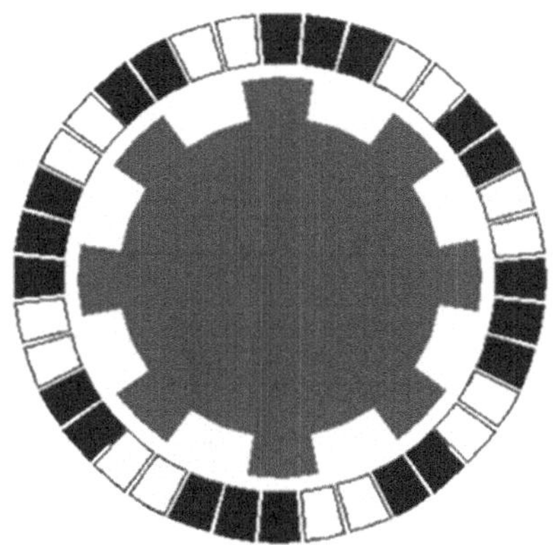

Fig. 4.68 Alternative motor design: small stator poles, big rotor poles

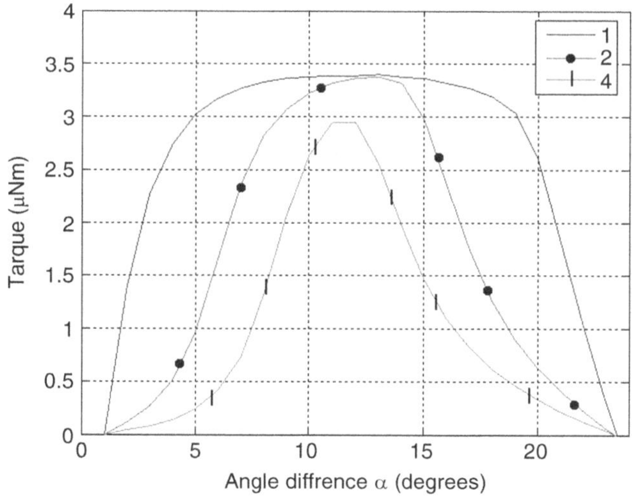

Fig. 4.69 Torque created by a rotor pole of a specified ratio between rotor and stator pole angular sizes

complicated than the one of a three-phase micromotor, because the necessary number of phases is at least equal to ratio between stator and rotor poles.

Such motor can have more than one stator pole acting on a single rotor pole. But as simulated, the torque created by a group of adjacent stator poles is always smaller than the one of a single pole of an equivalent size (Fig. 4.70). Similarly as analyzed earlier, highest torque is created when the corresponding set of active stator poles is of equal angular size as the rotor pole. Stator poles must be insulated from one another, thus the torque created by them is always smaller than the ideal case. Since the number of active poles pairs is always higher, the torque of such motor will also be higher.

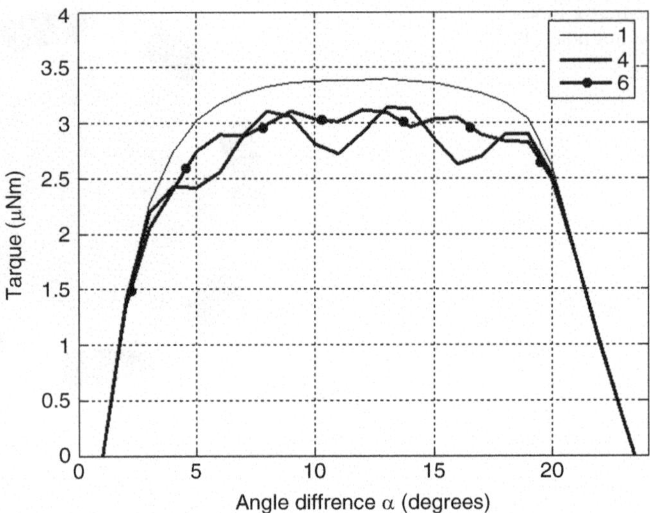

Fig. 4.70 Comparison of torques created by a single pole and a set of smaller poles having the same total angular size

The alternative design motor with many stator poles can create higher torque than a normal three-phase micromotor equipped with the same rotor. But such motor has unreasonably more complicated control due to a large number of necessary phases.

If exactly the same number of stator poles grouped in three phases is set and instead another rotor is used that is designed according to established guidelines, the torque will be similar, control simpler and production costs lower.

References

1. Lafontan X, Pressec F, Beaudoin F, Rigo S (2003) The advent of MEMS in space. Microelectron Reliab 43(7):1061–1083
2. Li G, Aluru NR (2002) A Lagrangian approach for electrostatic analysis of deformable conductors. J Microelectromech S 11(3):245–254
3. Zhang W-M, Meng G (2006) Friction and wear study of the hemispherical rotor bushing in a variable capacitance micromotor. Microsyst Technol 12(4):283–292
4. Yi F, Peng L, Zhang J, Han Y (2001) A new process to fabricate the electromagnetic stepping micromotor using LIGA process and surface sacrificial layer technology. Microsyst Technol 7(3):103–106
5. Preumont A (ed) (2002) Responsive systems for active vibration control. Kluwer, NATO Science Series
6. Behjat V, Vahedi A (2006) Minimizing the torque ripple of variable capacitance electrostatic micromotors. J Electrostat 64(6):361–367
7. Bishop D, Gammel P, Giles CR (2001) The little machines that are making it big. Phys Today 54(10):38–44

8. Stark B (1999) MEMS reliability assurance guidelines for space applications. JPL NASA, Pasadena, CA, pp 99–1
9. Hameyer K, Belmans R (1999) Design of very small electromagnetic and electrostatic micro motors. IEEE T Energy Conver 14(4):1241–1246
10. Maenaka K, Ioku S, Sawai N, Fujita T, Takayama Y (2005) Design, fabrication and operation of MEMS gimbal gyroscope. Sensor Actuat A Phys 121(1):6–15
11. Yang J, Ono T, Esashi M (2002) Energy dissipation in submicrometer thick single-crystal silicon cantilevers. J Microelectromech S 11(6):775–783
12. Vincent J, Rider P, Phan-Thien N (1998) A boundary element analysis of the fluid flow around micromotors. Eng Anal Bound Elem 21(3):277–284

Chapter 5
Technological Realization of MEMS Structures and Their Experimental Investigation

Abstract The last chapter presents the developed surface-micromachining technology that was patented by the authors and is suitable for fabrication of various MEMS actuators and sensors. Peculiarities of the microtechnology are highlighted such as utilization of special fractal microstructures, which enhance bonding strength of the microdevice to a substrate resulting in increased reliability. The chapter also describes design issues of electrostatic microswitches that were fabricated using the developed process and may be successfully implemented into various "system-on-chip" applications. Results of electrical and dynamic testing of the fabricated microswitches are provided including measurements of actuation voltages and natural frequencies with mode shapes. Microscope-based probe station and laser Doppler vibrometry system used for experiments are described as well. This chapter also provides overview of the developed micromotor fabrication technology that is based on application of standard UV lithography and plating. Produced prototypes of the electrostatic micromotors are demonstrated including their specifications. Lastly, a device developed for the micromotor electric control is presented that is able to adjust voltage, frequency, number of phases and other parameters.

The capabilities of the current photolithography processes and microfabrication techniques are inadequate compared to the requirements for production of high – performance actuators, such as microswitches and micromotors. The resulting inherent imperfections in the mechanical structure significantly limit the performance, stability and lifetime of these devices. For this reason, fabrication and commercialization of high performance and reliable MEMS devices have proven to be extremely challenging. Inevitable fabrication imperfections affect both the geometry and the material properties of MEMS structures. The dynamical system characteristics are observed to deviate drastically from the designed values and also from device to device, due to slight variations in photolithography steps, etching processes, residual stresses and construction errors.

Consequently, the current state of the art micromachined devices such as microswitches and micromotors require an order of magnitude improvement in their models, construction and manufacturing process in order to achieve better performance and lifetime. Fabrication imperfections and variations during the operation time of these

V. Ostasevicius and R. Dauksevicius, *Microsystems Dynamics*,
Intelligent Systems, Control and Automation: Science and Engineering 44,
DOI 10.1007/978-90-481-9701-9_5, © Springer Science+Business Media B.V. 2011

devices introduce significant errors, which have to be compensated by advanced control architectures and limitations of the existing micromachined structures have to be eliminated by modeling and analyzing them before any production steps.

5.1 Fabrication of Micromotor Prototypes

The investigated design concepts were implemented in an in-house micromachining process using only locally available equipment. The utilized fabrication processes and related equipment are presented in the following sections.

There are two main production processes of devices such as micromotors described in scientific articles: surface micromachining and LIGA. The biggest disadvantage of surface micromachining is that only very thin, and thus, very low torque motors can be fabricated. LIGA, on the other hand, enables production of much higher devices, and thus, higher torque motors, but is very expensive and available only in institutions that have that kind of equipment.

For these reasons, a three-mask process based on LIGA technology was developed [1–5]. It relies on conventional ultraviolet (UV) photolithography together with electrodeposition, which is similar to LIGA process but does not need expensive X-ray lithography. This technology is much cheaper, but is applicable only for layer thicknesses up to 30 μm.

The thickness of motor was selected to be 10–30 μm, and UV photolithography technology together with thick photoresist and electroplating was used. The positive tone photoresist ma-P 100 from Micro Resist Technology GmbH was critical element in producing micromotors. It is one of few UV photoresists capable to produce thicknesses higher than 10 μm. The "ma-P 100" is highly viscous tone photoresist with major components Naphtochinondiazid, Novolak and safer solvents dedicated to be used in microelectronics and micromechanics. Theoretically the photoresist demonstrates high stability in acidic and alkaline plating baths for film thicknesses up to 60 μm.

A selection of electroplating process and its material were also considered. A high modulus of elasticity and elastic behavior over a wide mechanical strain range is required for mechanical movable structures. Among many possibilities, nickel was selected because of its compatibility with in-house equipment and good mechanical properties.

Photolithographic masks presented in Fig. 5.1 were designed for production of the first prototypes. The rotor had 290 μm diameter, axle – 140 μm, the outside diameter of stator poles – 440 μm. Overall area needed for the motor was 1,200 × 1,300 μm. The rotor had six poles and the stator – seven poles. Together with the ground connection it made up eight connection points.

This design required a rather simple switching sequence that was good for manual testing but not appropriate for running motor in automatic electronically controlled mode. Simplicity in manual switching of such motor is because of one pole difference between rotor and stator. If stator poles are switched one after

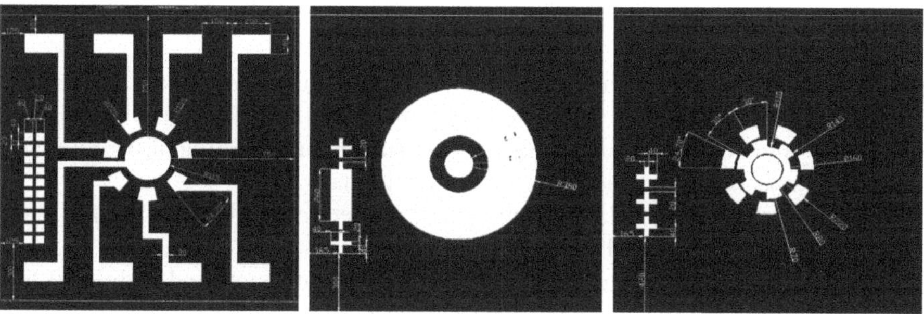

Fig. 5.1 Photolithographic masks for the first prototypes used in: step 4 (*left*); step 7 (*middle*); step 10 (*right*)

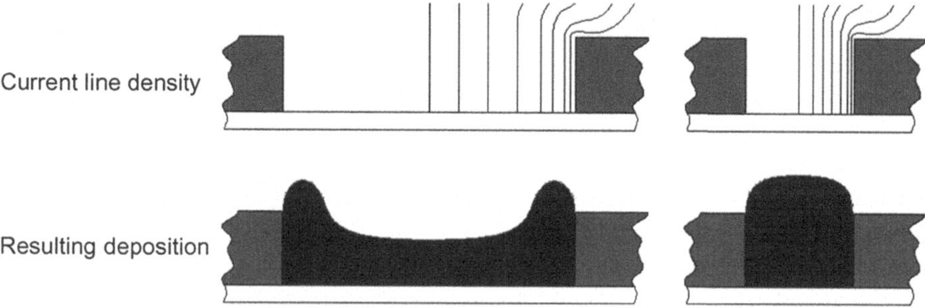

Fig. 5.2 Steps of micromotor production process

another in clockwise direction, the rotor would rotate counter-clockwise and vice versa. seven poles were too many for creation of seven phase control system, though the most important task regarding the first prototypes was establishment, verification and further development of the technology and this task was accomplished successfully.

The micromotors were produced using the following procedure (Fig. 5.2):

1. The process starts with preparation of special ST-32-1 crystallized glass ceramic substrates. The substrates are chemically cleaned in organic solutions and later in plasma of O_2/N_2 gases.
2. A layer of thin photoresist is spun on the substrate and cured in an oven.
3. Chrome layer of about 30 nm and further gold layer of about 200 nm thicknesses are deposited using electron beam evaporation technique.
4. Basic patterning of rotor and stator electrodes is performed using lift-off lithography using first photolithographic mask.
5. Electron beam evaporation is performed to deposit a sacrificial copper layer of 300 nm thickness. The layer covers the whole area of the substrate.
6. A layer of thin photoresist is spun on the substrate and cured in an oven.
7. A patterning of the copper layer to define area of the free rotating rotor is performed using the second mask.

8. The copper layer is etched away using H_2SO_4:CrO_3:H_2O etchant to uncover free rotating rotor area.
9. A thick positive tone photoresist ma-P 100 (provided by Micro Resist Technology GmbH) layer is spun and cured in an oven. Thickness is 10–30 μm.
10. A patterning of photoresist using UV exposure and uncovering of the rotor and stator areas using the third photolithographic mask.
11. Nickel layer is electroplated from sulfamate electrolyte ($Ni(NH_2SO_3)_2$:$4H_2O$) fabricating the rotor and stator structures.
12. The thick photoresist film is removed away using etchants and additional full UV exposure.
13. Finally, using a selective wet copper etchant the sacrificial layer is partly removed revealing the rotor.

The crystallized glass ceramic material is originated from silicon glass, though differs in its crystal structure which is more similar to ceramics, has a little bit bigger and more packed crystals making the substrate extremely non-porous (Fig. 5.3).

Thicker photoresist described in step 9 has different mechanical properties, and requires spinning speeds to be slower.

The thickness of the deposited nickel film depends on the thickness of photoresist and has to be controlled to be some micrometers thinner. A small area of photoresist above a substrate needs to be uncovered to connect a negative electrode of a circuit. Thus these areas become cathode. When a current is switched on, the electrolysis process begins and nickel ions flow onto uncovered copper of the substrate. If mask is very thin, because of current crowding large areas would grow with a raised rim near the interface with the resist (Fig. 5.4). In cases where resist is thick by comparison to the dimensions of the pattern, the current crowding is not as pronounced though it still exists, so plating has to be stopped at ~2/3 of total depth.

If the sacrificial copper layer is fully etched away, the rotors will be released and flow away into liquid etchant because they are not fixed from the top. For this reason, care has to be taken to keep the rotors on the substrate by etching the layer not completely through. Later a gentle touch by a needle is enough to let the rotors go.

5.2 Features of Produced Micromotors

Many variations of micromotors were manufactured for evaluation of technology (Fig. 5.5). The major differences among them were thicknesses. UV photolithography, unlike X-Ray lithography has a rather low aspect ratio, and all patterned features have wall inclination of about 4°. The 15:1 height-to-width ratio has minor influence on outside contours of the motor but is very important for narrow gaps between features. For this reason the projected gap between rotor and its axle on the mask was 10 μm. For thicknesses up to 30 μm the gap is wide enough to separate rotor from its axle, but if higher depositions of nickel are done, the rotor will remain in touch with its axle (Fig. 5.6).

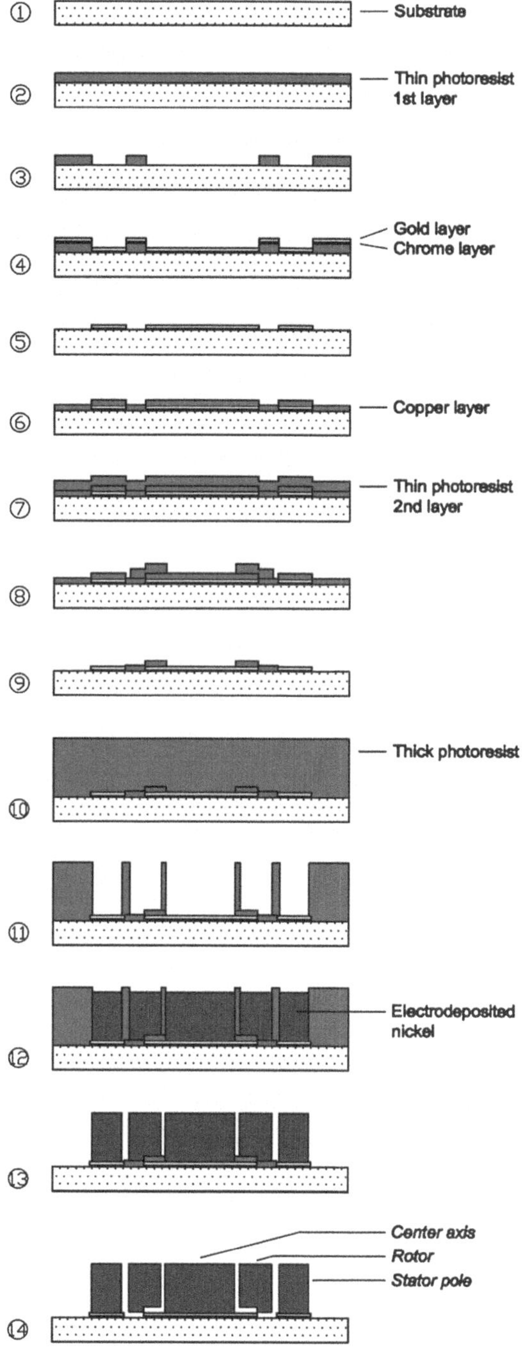

Fig. 5.3 ST-32-1 ceramic glass substrates

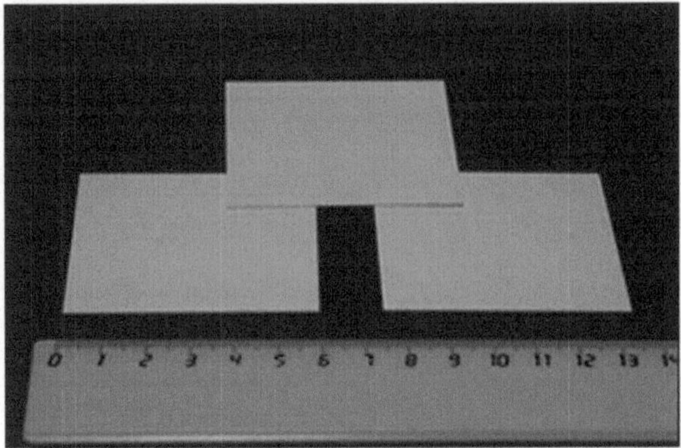

Fig. 5.4 Representation of current lines that create raised rims during electrodeposition: *Left*: large areas create only raised rims. *Right*: small areas create rounded constructions

Fig. 5.5 Image of the developed micromotor

Fig. 5.6 SEM image of the fabricated microrotor with stuck axle positioned on the top of a needle

Following success of the first prototypes, after extensive research and development, new designs of photolithographic masks for micromotors according to results of geometrical and electrostatic analysis were designed (Fig. 5.7).

As concluded earlier, rotors of micromotors have to be loaded by electrostatic forces symmetrically, i.e. forces from one side of rotor have to have counterbalance forces on the other side. For this reason, masks included 5 different micromotor designs having different number of rotor and stator poles (Fig. 5.8 and Table 5.1). All motors have four contact points: three for connecting control using three phases and one for grounding stator. The first two rotors have three stator poles and correspondingly four and five rotor poles. They are not axisymmetrically loaded but were selected due to their simple design. The other motors are loaded axisymmetrically. The depicted third and fourth micromotors have six stator poles and the last one has nine rotor poles. Correspondingly they have four, eight and six rotor poles. A disadvantage is that is not possible to connect three phases on one layer, thus additional layer or connecting leads need to be used. The 6-stator pole motor needs two interconnecting leads and the 9-pole motor needs even four interconnecting leads (Fig. 5.9 and Fig. 4.2).

The new production technology enables production of micromotors 2–5 times cheaper and using standard semiconductor UV fabrication equipment. The resolution and sizes of structures is the same as of integrated circuits. Diameters of motors can be up to 2,000 μm and smallest features are of couple micrometers size (Fig. 5.10–5.13). Also other materials can be selected for electroplating, so motors can be produced not only out of nickel, but aluminum and some others. The technology as most other MEMS technologies do not allow fabrication of rotor and axle of different materials, thus the typical tribological micromotor problems remain. The devices experience high stiction forces and can hardly spin.

Stiction can be prevented or reduced using several drying techniques available, like supercritical drying and sublimation drying to avoid the liquid–solid interface, therefore no surface tension induced capillary force will develop during drying.

A typical sublimation process starts from immersing the polymer structures in a small Isopropyl Alcohol (IPA) bath to remove the un-polymerized resin. It is followed by replacing IPA by dispensing t-butyl alcohol into the immersed sample and keeping the bath with constant volume of liquid solvent. After all solvent is replaced in the bath, the wet sample, together with the solution and container, is placed on Peltier chip and refrigeration is started. When the liquid solvent is totally frozen, pumping is started till all t-butyl alcohol solid in the container disappears [6]. Structures that do not experience capillary forces do not get stuck.

5.3 Micromotor Control Device

Using the MEMS manufacturing technologies above, or variations of these, it is possible to create a variety of microscopic actuators. However, simply using these techniques to fabricate components does not realize the full potential of MEMS. The

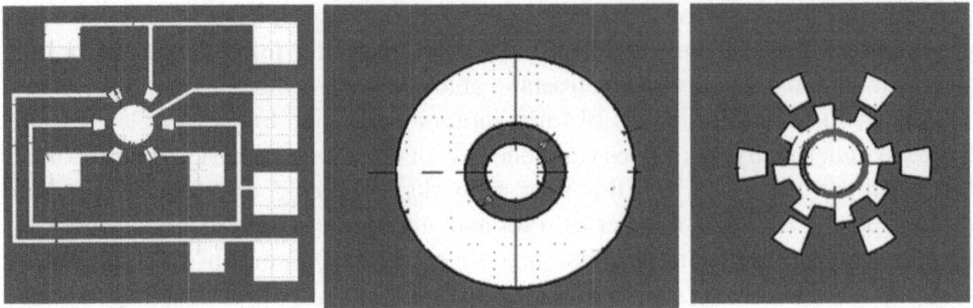

Fig. 5.7 Photolithographic masks for second generation micromotors used in: production step 4 (*left*); production step 7 (*middle*); production step 10 (*right*)

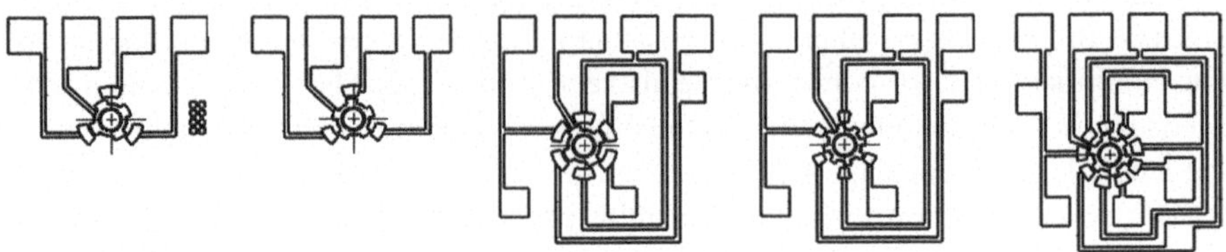

Fig. 5.8 Developed micromotor designs. *From left to right*: 3-stator 4-rotor pole; 3-stator 5-rotor pole; 6-stator 4-rotor pole; 6-stator 8-rotor pole; 9-stator 6-rotor pole

Table 5.1 Comparison of developed micromotor designs

Stator poles	3	3	6	6	9	12	15
Rotor poles	4	5	4	8	6	8	10
Channels (phases)	3	3	3	3	3	3	3
Relative torque	4.4	4.5	8.9	8.2	13.1	16.4	19.3
Poles per channel	1	1	2	2	3	4	5
Torque per pole	4.45	4.5	4.45	4.1	4.38	4.1	3.85

cost of the MEMS is often dominated by the cost associated with control, packaging and assembling the individual components. A basic MEMS actuator might only cost a few Euros, but the cost of a useful sensor or control system incorporating that actuator might be hundreds or thousands more. The true power and impact of microscopic devices comes when MEMS and CMOS control electronics are on the same piece of silicon. Since it is possible to batch fabricate the entire system on a single chip, the costs associated with packaging and assembly are greatly reduced, leading to extremely powerful, low cost solutions for many sensing applications.

The manufacturing process of the designed micromotor as many other MEMS devices is compatible with IC technology. This means that its control electronics for the motor can be integrated on the same chip using microcontroller and some switches or transistors. Alternatively, MEMS switches can be incorporated here, too.

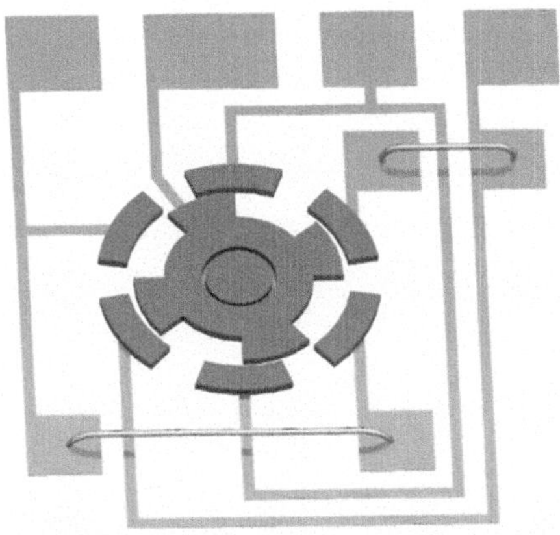

Fig. 5.9 Schematic drawing of connecting leads in an axisymmetrically loaded micromotor

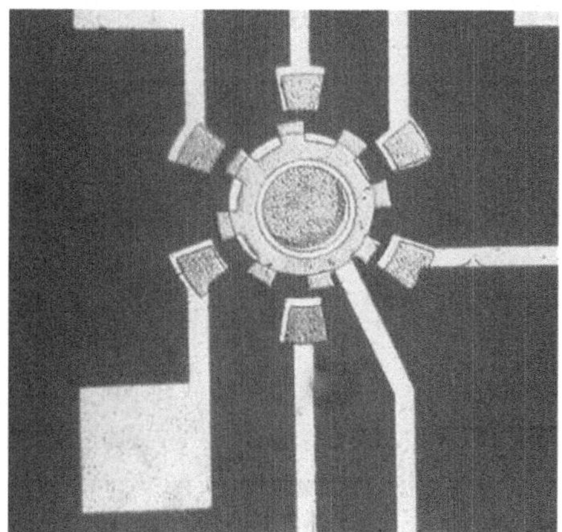

Fig. 5.10 8-pole axisymmetrically loaded rotor and 6-pole stator motor

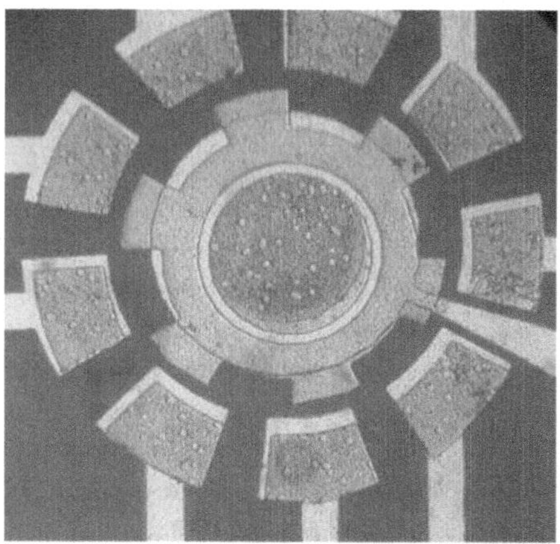

Fig. 5.11 9-pole stator motor

Fig. 5.12 Good mask alignment and thin rotor stuck to surface of crystallized glass plate

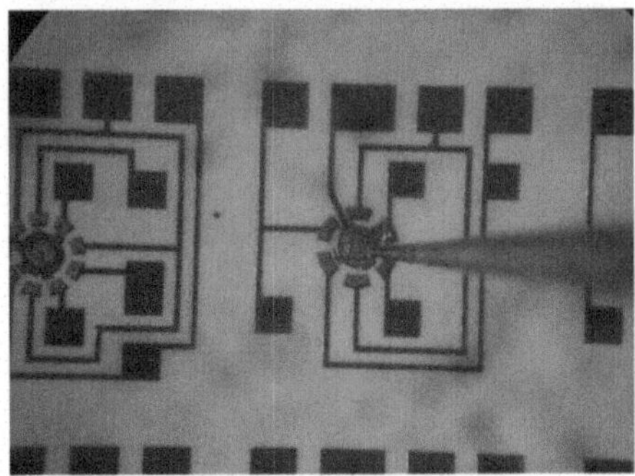

Fig. 5.13 View of multiple fabricated motors

To control and analyze prototype micromotors a stand-alone control unit was necessary to be built that would have most user interface elements according to output results of the MATLAB program.

The control unit named MVV-1 (Figs. 5.14 and 5.15) has three phase outputs and a ground connector that are intended to be connected to micromotor poles. Motor can be operated clockwise and anticlockwise. Frequency of switching in auto mode is up to 1 kHz but the maximum speed of motor depends on the number of rotor poles the motor has. A 4-pole rotor spins 1/4 of a circle during full three phase cycle. A 5-pole rotor spins 1/5 of circle during the same interval. Thus the maximum speed of a 4-rotor-pole motor at 1 kHz switching frequency is 15,000 rotations per minute (rpm). Consequently, the 5-rotor-pole motor has maximum speed of 12,000 rpm at 1 kHz.

The motor can be switched to a manual mode and run one pulse at a time. By pressing a button a single pulse on one of phases is outputted and rotor rotated by one step. This mode is useful for detailed analysis of every step of the motor just like single steps of stepper motors.

If switching is started to a motor at full speed right from still state (speed – 0 rpm), it is highly possible that the motor doesn't start spinning due to its inertia. Even if it started, the transient process would still be not scientifically recurrent and would

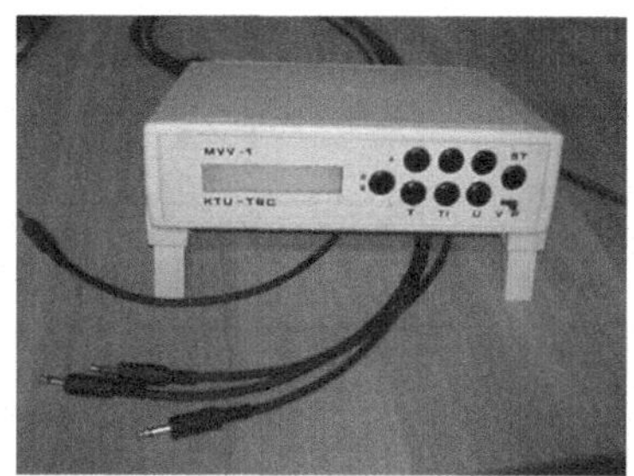

Fig. 5.14 Micromotor control device

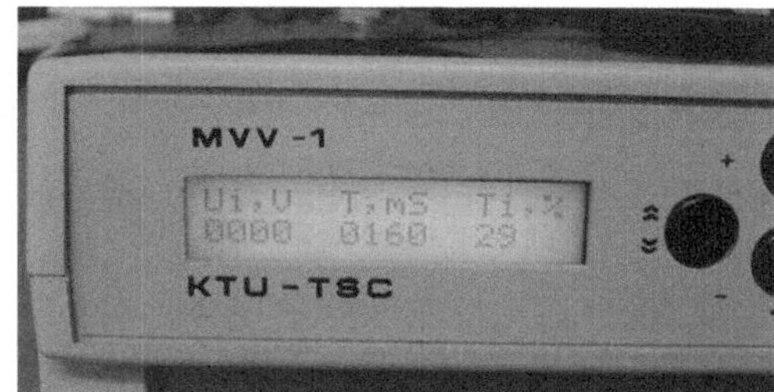

Fig. 5.15 Screen of the device

definitely cause vibrations of the rotor. Thus, special start-up transient switching sequences were developed, too. In a controlled adjustable time interval the motor is spun from still to its full speed. This is done by entering start-up speed, final speed, the time difference between them and type of curve: linear, parabolic, logarithmic or other. This way it is possible to indirectly assess inertia properties of individual rotors.

There is also an RS 232 port connection to a computer. This enables loading any switching sequences for the motor for complex transient measurements.

Potential difference is adjustable electronically from 0 to 250 V. Higher potential produces higher force momentum. For this reason, a wide range of voltages was selected because friction of motors is high and process-dependent. Adjusting operating voltage enables measurement of friction of individual rotors.

Duty cycle of phases is also adjustable from 0.05 to 0.95 of full period. This parameter has an important influence on the speed of the rotor, and ideally should be equal to the specified in the simulation.

The final adjustable parameter is the phase shift. The first phase is not adjustable, but the second and third phases are equally adjusted $\pm 60^\circ$ regarding their default 120° and 240° values. If the second phase is adjusted to 20°, then the third phase is set to 40°, Fig. 5.16. The first phase is always at 0°.

Using the listed range of adjustable parameters, most control situations of the motors can be examined experimentally (Table 5.2).

The control unit can be powered through outlet, or can be used as a portable device with a 9 V battery. Impulses from the control unit are transmitted to micromotor through a microscope station by special probes (Fig. 5.17 and 5.18).

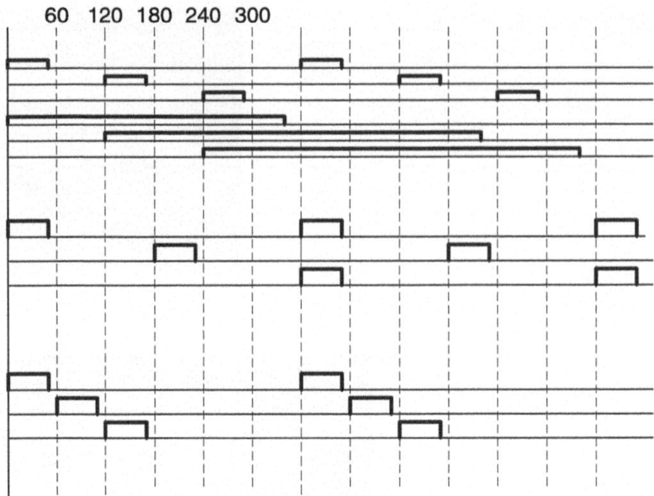

Fig. 5.16 Duty-cycle and phase shift pulse sequences supplied to a micromotor

Table 5.2 Parameters of micromotor control unit

Parameter, units	Values	Notes
Frequency	(0–1) kHz	
Duty cycle	(0.05 T – 0.95 T)	T – period of signal
Phase shift	120° ± 120°	Equal phase shift for second and third channel
Amplitude (voltage)	(0–250) V	

Using standard in-house IC fabrication equipment a new, cheaper process for manufacturing micromotors was developed as compared to common LIGA technology. Two sets of prototypes were produced using three sets of masks, standard and thick photoresist layers, sacrificial copper layer and nickel electrodeposition. Thickness of produced motors was 10–30 μm. The devices experienced high friction forces.

A special fully programmable control device was produced for phase switching. The device generates up to 1 kHz pulses on three phases that enables to spin a motor up to 15,000 rpm depending on configuration of a motor.

5.4 Design of Electrostatic Microswitches

Electrostatically-actuated microswitch (Fig. 5.19) was developed by the co-authors [7]. Original surface micromachining technology that was used for fabrication of this microdevice was patented [8]. This microfabrication technology is suitable for fabrication of various MEMS actuators and sensors. It is compatible with manufacturing methods that are used in microelectronics therefore the

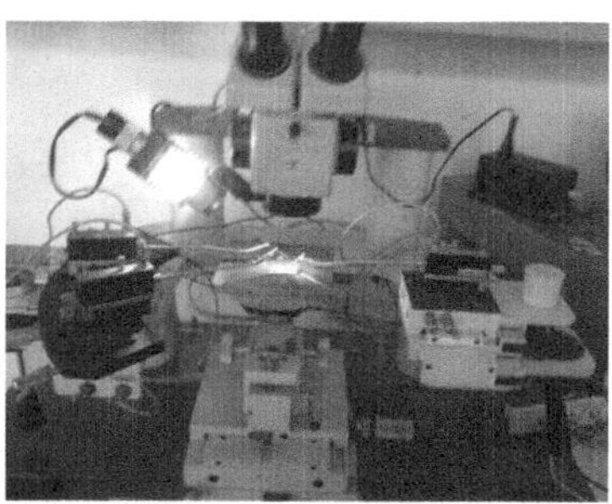
Fig. 5.17 Micromounting unit with probes

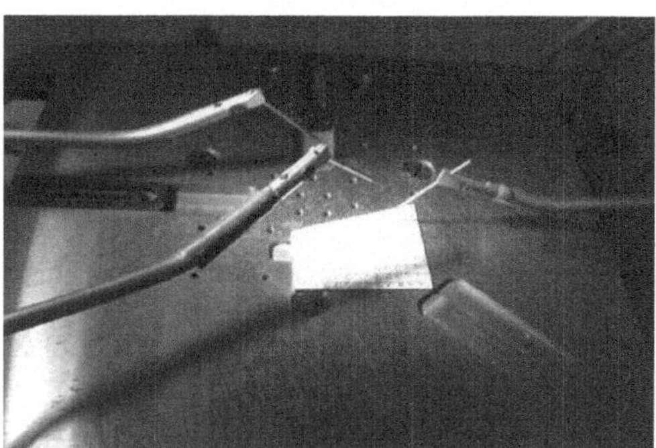
Fig. 5.18 Probes for connecting DC source to micromotor

microswitches may be integrated together with different electronic devices onto the same microchip for diverse "system-on-chip" applications. They are an attractive alternative to electromagnetic and solid-state switches and could be used in a variety of radio-frequency applications, including cell phones, phase shifters, and smart antennas, as well as in lower frequency applications, such as automated test equipment, industrial and medical instrumentation, etc. Introduction of special fractal microstructures constitutes the originality of the developed microtechnology since they enhance the bonding strength of the microdevice to the substrate thereby increasing its reliability.

Microswitch configuration and principle of operation. Configuration of the developed microswitch is presented in Fig. 5.20. It is a three-terminal device with base electrode (source) 5, actuation electrode (gate) 6 and contact electrode (drain) 7. Main structural element of the microswitch is a microcantilever 9, which base is attached to the substrate 1 in the anchor region 2 and is electrically connected to the source electrode (Fig. 5.21 (bottom)). The microcantilever is bimetal and made of layers of nickel 9 and gold 10 with total thickness in the range of 1.5 ÷ 3.0 μm. The thickness of gold film is approximately 0.2 μm and it serves to reduce contact resistance. The width of the microcantilever is 30 μm and length ranges from 67 to 117 μm. When the voltage is applied between the source and gate electrodes, the microcantilever is pulled down by the electrostatic forces until its free end contacts the drain and completes an electrical path between the source and the drain allowing

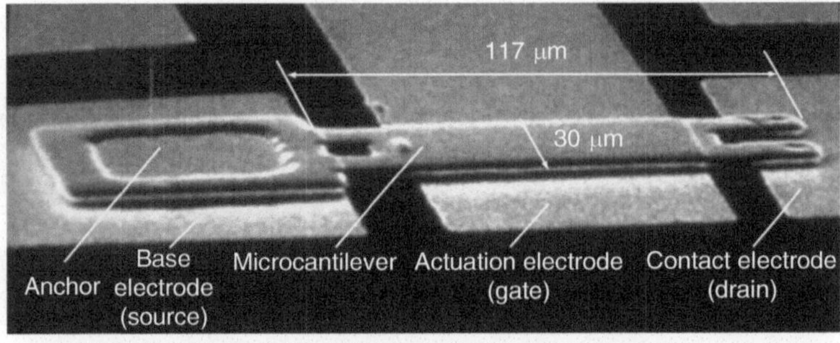

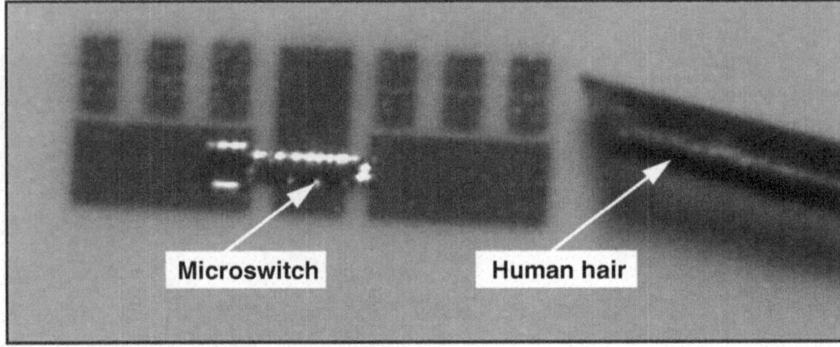

Fig. 5.19 *Top*: scanning electron microscope (SEM) image of the developed surface-micromachined electrostatically actuated microswitch. *Bottom*: general view of the microswitch in comparison to the size of a human hair

electrical current to flow. When the voltage is removed, the elastic restoring force of the microcantilever returns it to the original position.

Essential features of the microswitch:

1. It is a resistive-type switch, i.e. it uses metal-to-metal direct contact principle in order to perform switching function in the circuit.
2. The microdevice has been fabricated on substrates (wafers) made of: (a) high-resistivity semiconductor (silicon), which enables integration of microswitches into monolithic integrated circuits, and (b) dielectric materials, namely, quartz and sital (cervit, zerodur), which is a type of glass-ceramic characterized by strong adhesion, high surface quality and nearly zero coefficient of thermal expansion. This allows the microswitches to be added into hybrid circuits.
   Substrates contained an array of fabricated microswitches having different geometric dimensions, e.g. length, thickness or gap size.
3. Novel technological solutions were implemented in the microfabrication process – special fractal microstructures were formed in the anchor region in order to increase bonding strength of the microswitches to a substrate and thereby enhance their durability.
4. Microcantilever has got two miniature protrusions (contact tips 8) on the lower side of the free end, which are small by design ($4 \times 4\ \mu m^2$) and their function is

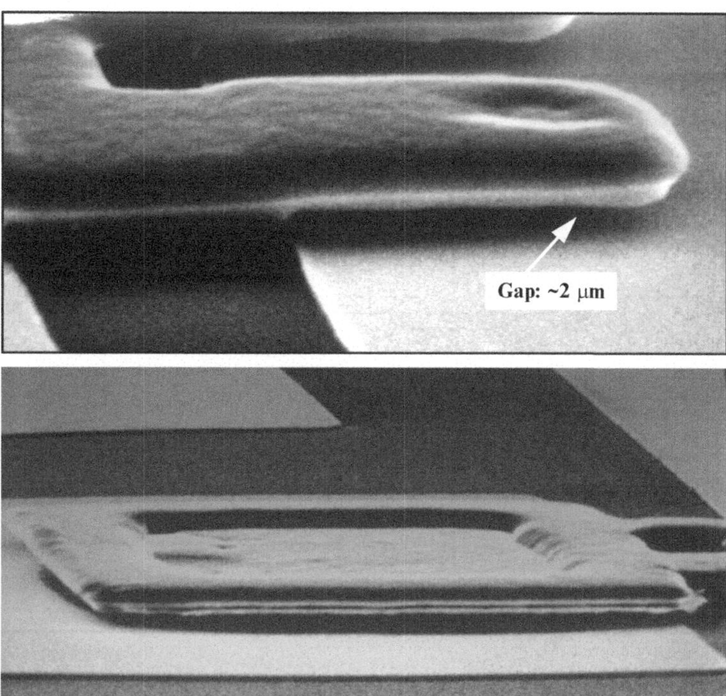

Fig. 5.20 SEM images. *Top*: fork-type free end of the microcantilever shown in detail. *Bottom*: microswitch anchor (support) region, where the microcantilever is fixed to a substrate. The characteristic feature of the anchor is its "cup-shaped" profile, which results in non-ideal cantilever-type fixing conditions

to: (a) increase the pressure of contact thereby ensuring more reliable contact and device functioning; (b) to control actuation hysteresis generated due to the pull-in instability phenomenon since the spacing between the tips and the drain can be varied during fabrication process (for the fabricated microswitches this distance is in the range of $1.0 \div 2.0$ μm). In this way non-hysteretic microswitch may be developed, in which collapsing (pull-in) action is not manifested during microcantilever deflection (i.e. contact is achieved before pull-in instability point) resulting is lower switching time and pull-in voltage.

5. Figure 5.21 (top) illustrates in detail that free end of the microcantilever is fork-shaped. This is meant to reduce the area of the microcantilever that overlaps with the drain electrode. This, in turn, reduces the capacitance between the source and the drain, thereby increasing the breakdown (stand-off) voltage, i.e. the potential between the source and drain that is capable to close the switch (regardless of the potential between the gate and the source) and cause: (a) either excessive current to flow thereby destroying the device or (b) induce mechanical relaxation oscillations.

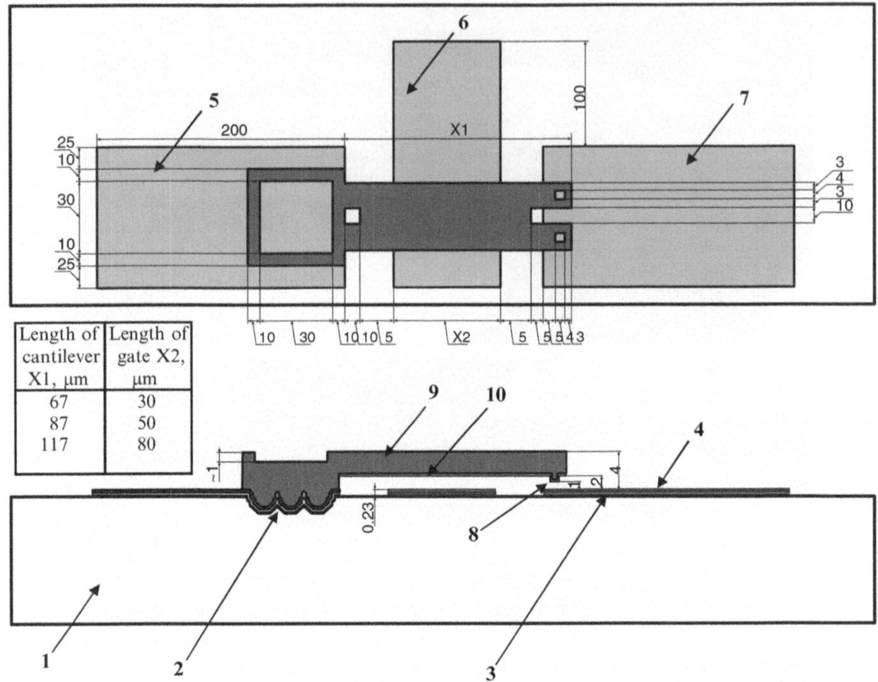

Length of cantilever X1, μm	Length of gate X2, μm
67	30
87	50
117	80

Fig. 5.21 Detailed schematic representation of the fabricated microswitch: 1 – substrate (Si, quartz, sital), 2 – anchor region with fractal microstructures, 3 – underlayer (Cr), 4 – contact layer (Au), 5 – base electrode (source), 6 – actuation electrode (gate), 7 – contact electrode (drain), 8 – contact tips, 9 – first beam layer (Ni), 10 – second beam layer (Au)

6. The microcantilever contains a hole near its fixed end. The fabrication technology, in general, allows producing microcantilevers with different size, number and location of holes. Their purpose is multiple: (a) to facilitate the escape of air from underneath the microcantilever thus controlling the level of air damping and consequently switching speed; (b) to improve the effectiveness of microcantilever release during fabrication process at etching stage by allowing etchants to better penetrate the area under the microcantilever through holes; (c) to control mass of the microstructure as well as its surface area thus affecting both switching speed and pull-in voltage.

5.5 Surface Micromachining Technology for Fabrication of Microswitches

Developed surface micromachining technology, which is used for fabrication of microswitches is an all-metal photolithography-based technology that uses a combination of different thin film deposition and etching processes such as wet and reactive ion etching, electron-beam vacuum evaporation, electroplating, etc.

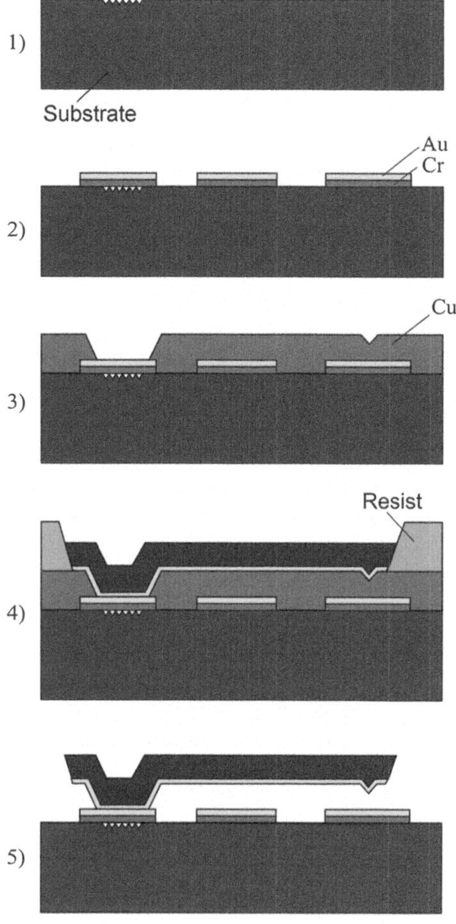

Fig. 5.22 Schematic representation of microswitch microfabrication sequence: 1 – patterning and etching of the microstructures in the source (anchor) area, 2 – "lift-off" photolithography of Cr-Au bilayer, 3 – processing of a sacrificial Cu layer, 4 – evaporation and "lift-off" of Au layer; patterning of the thick resist and Ni electroplating, 5 – removal of a sacrificial layer

The technological process consists of the following steps (Fig. 5.22):

1. Cleaning of substrate surface 1 chemically in organic solvents and by means of plasma of O_2 + N_2 gases. Afterwards, in order to increase strength of bonding of the microswitch to the substrate, special fractal microstructures of large surface area are formed in the anchor region 2. This accomplished by: (a) deposition of 100 nm aluminum layer by electron-beam vacuum evaporation; (b) etching of mask in the Al layer for the microstructures (etchant: Cr_2O_3:NH_4F:H_2O); (c) formation of fractal microstructures by combining wet and reactive ion etching processes. The microstructures are of stepped-pyramid shape and have high effective surface area, thereby ensuring strong attachment of microcantilever (to be formed later) to the substrate. Wet etching is carried out by using Sirtl etchant (HF:HNO_3:$C_2H_4O_2$), meanwhile reactive ion etching – in SF_6/N_2 high-frequency reactive gas plasma.

2. Further, the substrate is repeatedly cleaned chemically in organic solvents and in plasma of $O_2 + N_2$ gases. Then source 5, gate 6 and drain 7 electrodes are formed on the substrate. Topological pattern of the electrodes is obtained by means of so-called "lift-off" process: (a) photoresist is first deposited and patterned in order to define patterns of source 5, gate 6 and drain 7 electrodes; (b) then a layer of chromium (30 nm) 3 is vacuum deposited by electron beam and, subsequently, on top of it a layer of gold (200 nm) 4 is formed by thermoresistive vacuum deposition; (c) finally, the patterned photoresist with deposited metal layers atop is removed ("lifted-off") in organic solvent dimethylformamide (DMF).
3. Subsequently, a sacrificial layer of copper (3,000 nm) is deposited on the whole area of substrate by using electron beam evaporation. The copper is then photolithographically patterned twice. First of all, the copper layer is partially etched (etchant: H_2SO_4:CrO_3:H_2O) to define the contact tips 8 for the micro-cantilever. Etching duration directly determines the spacing between contact tips and drain electrode. During the second photolithography stage copper is etched completely away thereby uncovering a layer of gold in the anchor region.
4. Next fabrication step is formation of a cantilever beam 9. First of all a layer of gold (200 nm) 10 is formed by lift-off process, followed by photolithographical patterning and electrochemical deposition (electroplating) of layer of nickel (3,000 nm) by using sulphamate electrolyte $Ni(NH_2SO_3)_2$:$4H_2O$, pH = 3,6–4,2 and the following electroplating process parameters: cathodic current density – 5 A/dm^2, T = 45–60°C.
5. The formation of the switch is finalized by removing sacrificial layer of copper using selective wet etching (etchant: H_2SO_4:CrO_3:H_2O).

5.6 Electrical Probe Testing of Fabricated Microswitches

Objectives of the initial experimental investigation of fabricated microswitch samples were: (a) to check their functioning by inspecting them visually and performing resistance measurements for identification of defective devices and (b) to carry out electrical measurements of pull-in voltage of the microswitches.

Description of the experimental setup. Due to miniature size of the fabricated microdevices a special experimental setup with microscope and probe station was used. It consists of the following main components (Fig. 5.23 (top)):

1. Optical microscope 1 with 16× ocular and three selectable objectives of different magnification power (5×, 16× and 20×). The position of microscope may be changed along axes Z and Y.
2. Movable microscope stage 2 for putting test object. The stage may be positioned by using micrometric screws along all three axes. The position of the test object on the stage may be fixed by means of vacuum suction.
3. Three manually-adjustable probes 3 for connecting electrodes of the micro-switch to the electric measurement circuit. The position of probes may be changed by micrometric screws in vertical and lateral directions.

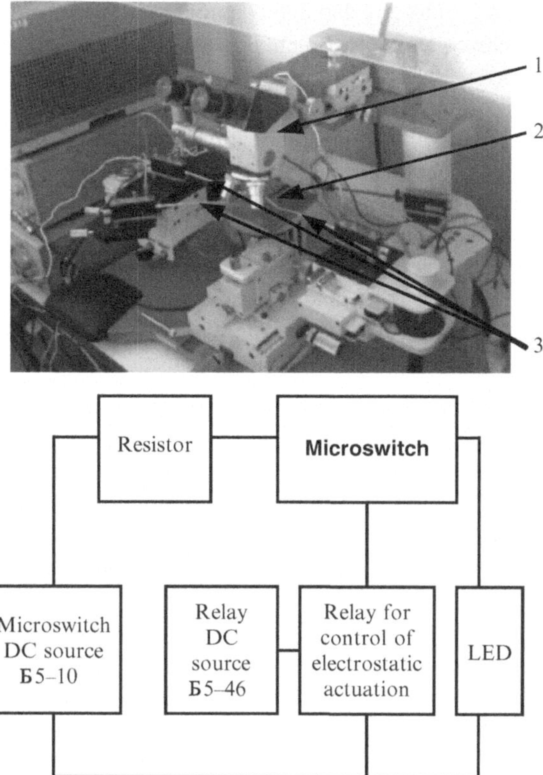

Fig. 5.23 *Top*: experimental setup for electrical probe testing of fabricated microswitches: 1 – microscope, 2 – XYZ-axis movable table for putting test object, 3 – probes for connecting test object to the electric measurement circuit. *Bottom*: block diagram of measurement set-up used for electrical characterization of fabricated microswitches

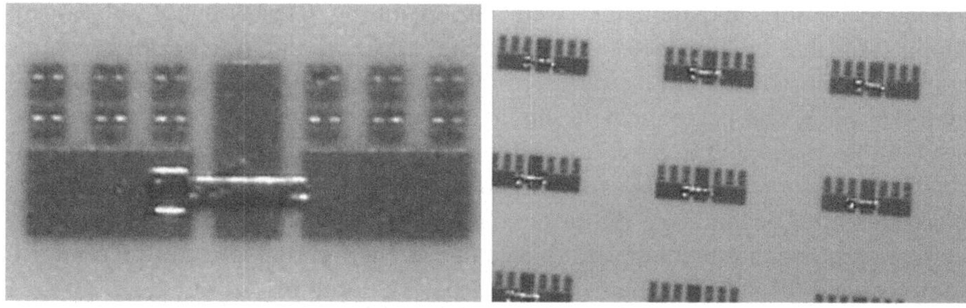

Fig. 5.24 Photos of fabricated microswitches

Visual inspection and resistance measurements. The first samples of microswitches were fabricated on three substrates made of different material: silicon, quartz and sital. Each substrate contained an array of approximately 560 (20 × 28) devices having microcantilevers of three different lengths: 67 μm, 87 and 117 μm (Fig. 5.24). First of all, fabricated microswitches were inspected visually one by one and this allowed locating of totally damaged devices that were not suitable for further testing. The aim of subsequent experimental work was to identify those microdevices that were defective due to either stiction (adhesion), which is a major problem in MEMS fabrication, and/or due to incomplete etching of sacrificial material underneath the microstructure during fabrication (remains of the sacrificial material in the gap can make the cantilever join with bottom electrodes). Gate and drain regions are those two possible places where microcantilever of the microswitch can adhere with bottom electrodes. Therefore measurements of electrical resistance were carried out between the source and the gate as well as between the source and the drain electrodes. The magnitude of resistance in those places where microcantilever was adhered to the electrode was about 4.0 ÷ 5.0 Ω. The results of the visual inspection and resistance measurements revealed the following:

1. Sital substrate contained the smallest number of defective microswitches.
2. In majority of cases the microcantilever was adhered to the drain electrode, which can be explained by the fact that in this region because of the protruded contact tips the spacing between the microcantilever and the bottom electrode is smaller in comparison to the gate region and this consequently facilitates the stiction.
3. The lowest percentage of defective microswitches was found among the shortest microdevices ($l = 67$ μm), which could be explained by: (a) higher stiffness of the shortest microcantilevers in comparison to the longer ones, which in turn impedes the stiction between contact tips and drain electrode, and/or (b) more effective etching of the sacrificial material underneath the shortest microcantilever (i.e. less remains of sacrificial material in the gap). It is obvious that it is easier to etch away the material underneath the shorter microcantilever than the longer one.

In order to reduce the level of defective microswitches due to stiction in drain region, technological parameters were varied during fabrication of subsequent samples of the microdevices. The spacing between the microcantilever tips and the drain was varied in the range 1.0 ÷ 2.0 μm and each time above-described control testing was performed. It should be noted that the results obtained and observations made during these tests allowed to reduce defectiveness of the microswitches, and served as a basis for subsequent improvement of developed surface-micromachining technology by introducing additional technological steps in the fabrication procedure that allowed to enhance strength of bonding between microcantilever and the substrate thus increasing device durability.

Pull-in voltage measurements. Electrical measurement circuit presented in Fig. 5.23 (bottom) was set up in order to perform electrostatic actuation experiments on microswitches and determine their pull-in voltage, i.e. the smallest actuation voltage at which the switch closes. For the actuation of the microswitch a DC power source Б5-10 was placed in series. Electromagnetic relay, powered by DC power source Б5-46, was inserted into the measurement circuit in order to have the possibility turn the actuation voltage on or off during testing. In order to limit electric current flowing through closed microswitch, 10 kΩ resistor was connected in series with the microswitch. Light-emitting diode (LED) was used in order to indicate the moment when the switch is closed. The pull-in voltage was measured by: (a) applying actuation voltage by turning on the electromagnetic relay, (b) increasing gradually actuation voltage with power source Б5-10 until LED flashes and recording at this moment indications of voltage with multi-meter UNI-T® M830B. Pull-in voltages were measured for microswitches of three different lengths: 67, 87 and 117 μm. The corresponding average values of the measured pull-in voltage are: 226.70, 128.40 and 59.90 V. Figure 5.25 shows measurement results presented in the form of histograms of the pull-in voltage variation for the tested microdevices. The histograms demonstrate fairly significant spreading of the results, which is a consequence of a number of different contributing factors such as variation in contact tip height, air gap spacing in the gate region and the microcantilever thickness. Microswitch pull-in voltage is particularly sensitive to the variation of the microcantilever thickness t since stiffness of cantilever is proportional to t^3. Therefore improved control of the aforementioned parameters during fabrication is required between substrates and across a substrate in order to reduce the observed variation of pull-in voltage.

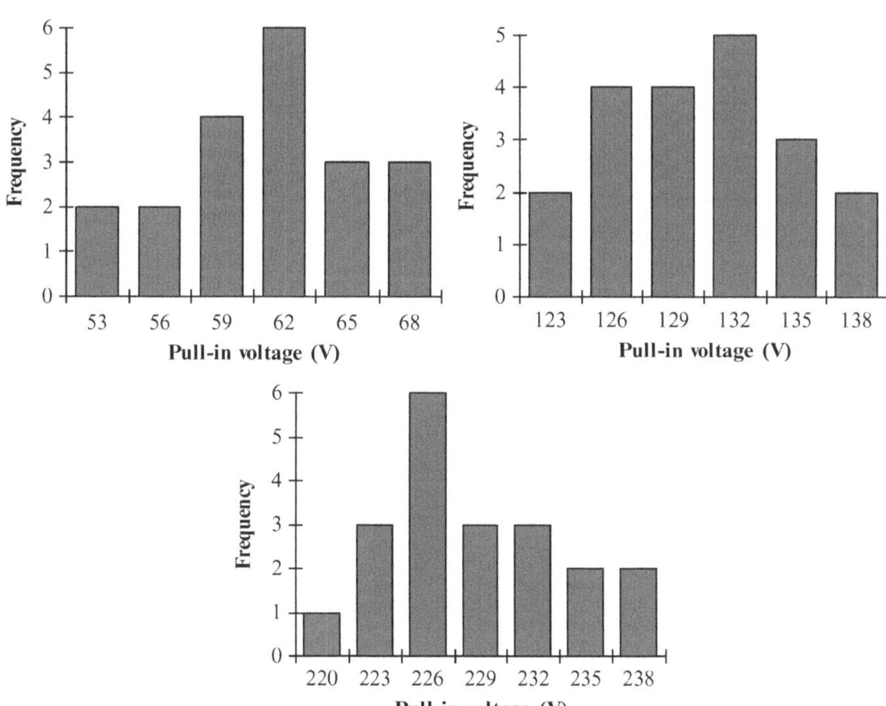

Fig. 5.25 Histograms of the measured pull-in voltage variation for the tested microswitches of different length. *Left*: $l = 117$ μm. *Right*: $l = 87$ μm. *Bottom*: $l = 67$ μm

5.7 Laser Measurements of Microcantilever Vibrations

Vibrometry basics. A laser Doppler Vibrometer (LDV) is based on the principle of the detection of the Doppler shift of coherent laser light that is scattered from a small area of the test object. The object scatters or reflects light from the laser beam and the Doppler frequency shift is used to measure the component of velocity which lies along the axis of the laser beam. Since laser light has a very high frequency Ω (approximately 4.74×10^{14} Hz), a direct demodulation of the light is not possible. An optical interferometer is therefore used to mix the scattered light coherently with a reference beam. The photo detector measures the intensity of the mixed light whose (beat) frequency is equal to the difference frequency between the reference and the measurement beam. Fundamental LDV configuration is similar to that of the Michelson interferometer as it is shown in Fig. 5.26 (left). A laser beam is divided at a beam splitter into a measurement beam and a reference beam which propagates in the arms of the interferometer. The distances the light travels between the beam splitter and each reflector are x_R and x_M for the reference mirror and vibrating object respectively. The corresponding optical phase of the beams in the interferometer is:

Reference:

$$\Phi_R = 2kx_R. \tag{5.1}$$

Measurement:

$$\Phi_M = 2kx_M, \tag{5.2}$$

where $k = 2\pi/\lambda$ and $\Phi(t) = \Phi_R$ - Φ_M.

The photo detector measures the time-dependant intensity $I(t)$ at the point where the measurement and reference beams interfere:

$$I(t) = I_R I_M R + 2K\sqrt{I_R I_M R}\cos(2\pi f_D t + \Phi(t)), \tag{5.3}$$

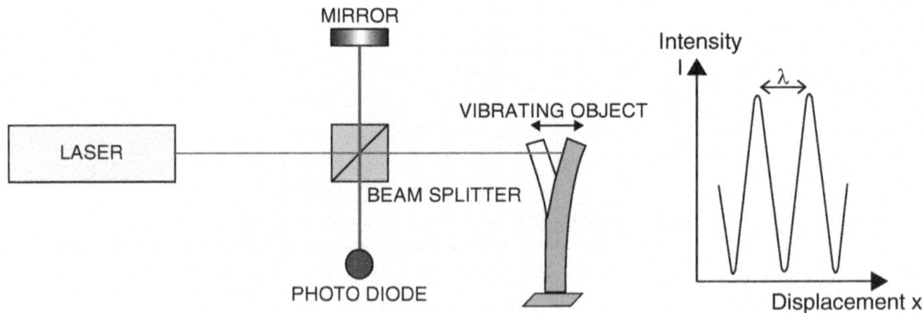

Fig. 5.26 *Left*: basic configuration of vibrometer. *Right*: changes of light intensity with respect to time

where I_R and I_M are the intensities of the reference and measurement beams respectively, K is a mixing efficiency coefficient and R is the effective reflectivity of the surface, f_D is the Doppler frequency generated by the vibrating object. Time varying interference phase information $\Phi(t) = 2\pi\Delta L/\lambda$, where ΔL is the vibrational displacement of the object and λ the wavelength of the laser light. If ΔL changes continuously the light intensity $I(t)$ varies in a periodic manner (Fig. 5.26 (right)). A phase change $\Phi(t)$ of 2π corresponds to a displacement ΔL of $\lambda/2$. The rate of change of phase $\Phi(t)$ is proportional to the rate of change of position which is the vibrational velocity v of the surface. This leads to the well known formula for the Doppler frequency f_D:

$$f_D = \frac{2v}{\lambda}, \tag{5.4}$$

where v has to be determined by solving the above Eq. (5.3) for $I(t)$ but it contains more variables than can be solved for with a single equation. To generate a solvable system of equations a frequency shift is added into one arm of the interferometer with the use of an acousto-optic modulator (Bragg cell). This frequency shift, f_B, is 40 MHz or higher [9]. Then the equation is modified as follows:

$$I(t) = I_R I_M R + 2K\sqrt{I_R I_M R}\cos(2\pi(f_B - f_D)\,t + \Phi(t)). \tag{5.5}$$

The Doppler frequency, and thus the object velocity, is now carried in the ~40 MHz signal and can be decoded using temporal heterodyne techniques.

Figure 5.27 (top) shows schematic block representation of the experimental setup that was used for dynamic characterization of fabricated microswitches. It consists of the following major components:

- Laser measurement sub-system: Polytec LDV system MSV-400 shown in Fig. 5.27 (bottom);
- Excitation sub-system: signal generator, amplifier, excitation piezoactuators;
- Data management sub-system: digital oscilloscope, computer with software for MSV-400 control, data processing and analysis.

Laser measurement sub-system. Main component of the experimental setup is Polytec microscope-based LDV system MSV-400, with essential features and technical specifications indicated in Table 5.3. MSV-400 includes the following modules (Fig. 5.28):

Fiber-optic vibrometer OFV-512
Vibrometer controller OFV-5000
Scanner controller (junction box) MSA-E-400
Nikon microscope with:

(a) Remote-controlled microscope adapter MSV-100 for coupling laser beam into microscope and beam scanning
(b) Digital progressive scan camera VCT-101

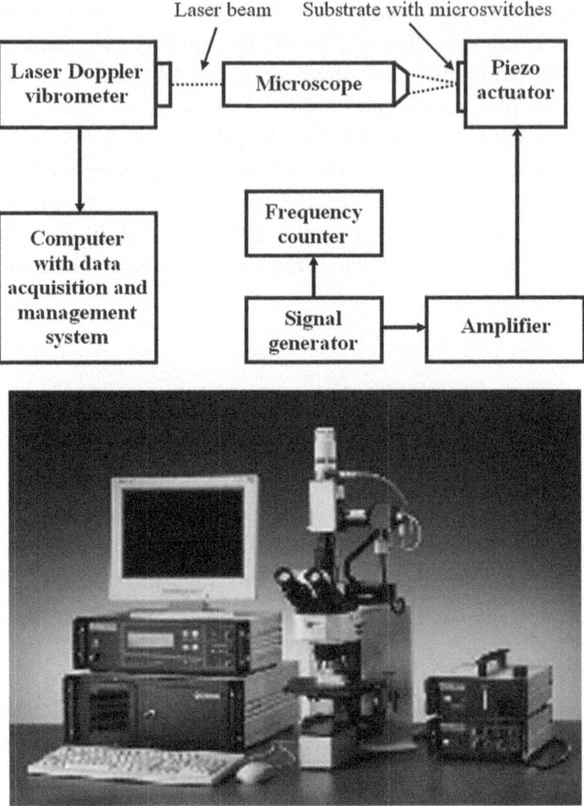

Fig. 5.27 *Top*: block representation of experimental setup used for vibration measurements of microcantilevers with laser Doppler vibrometer. *Bottom*: photo of Polytec LDV system MSV-400

Table 5.3 Summary of essential features and technical specifications of Polytec microscope-based laser Doppler vibrometry system MSV-400

Essential features	Technical specifications
Wide applicability and high versatility – measurement of out-of-plane velocities and displacements of both macro- and micro-structures. Measures periodic, random or transient motions	Measured vibration frequency range: 0.5 Hz ÷ 1.5 MHz Measured vibration velocity: 0.05 ÷ 10 m/s Measured displacements: from several nanometers to several centimeters
Non-contact, zero mass-loading on tested object, eye-safe laser	
Modular design, capability of differential measurements	Maximum displacement resolution: several picometers (depending on frequency)
High dynamic range, wide bandwidth, outstanding measurement accuracy and resolution, small spatial measurement spot	Maximum velocity resolution: 0.15 μm/s Smallest laser spot size: ~1 μm Size of scan area: ~3.1 mm^2 ÷ ~70 μm^2 with microscope objectives 4× and 100× respectively

Microscope adapter MSV–100

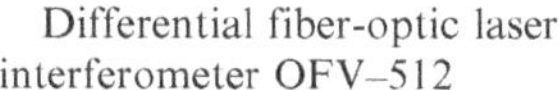

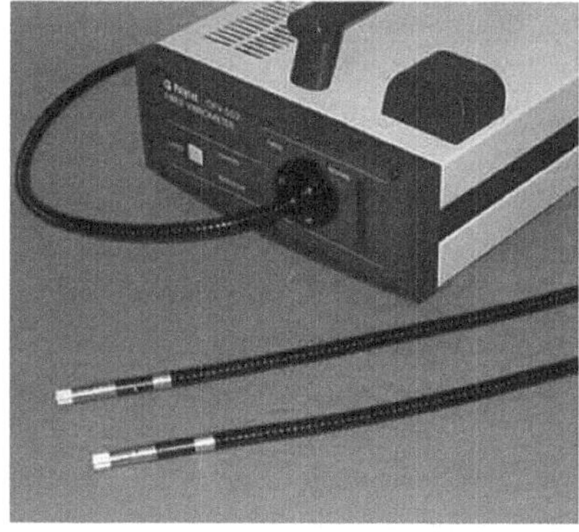

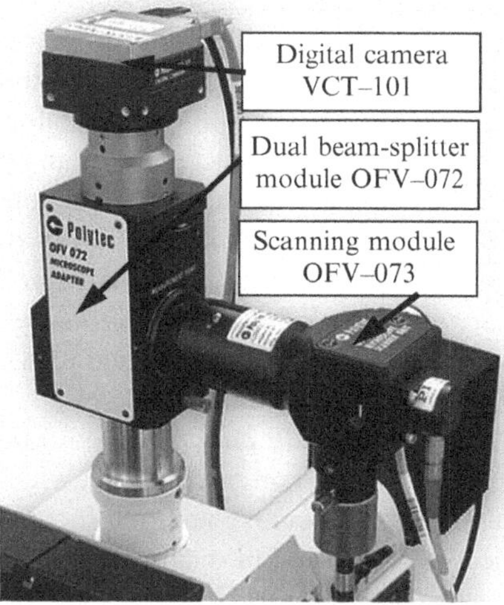

Scanner controller MSA–E–400

Vibrometer controller OFV–5000

Fig. 5.28 Components of Polytec laser Doppler vibrometry system MSV-400

Software for system control

Fiber-optic vibrometer OFV-512 is a two-arm differential single-point fiber-optic laser interferometer, which uses HeNe red laser as a coherent light source and permits direct measurement of differential movements between two monitored points. It utilizes flexible optical cable terminated with mini-lens, which allow to direct and focus laser beam onto the desired measurement point and to collect the reflected light. High-efficiency delivery and collection optics produce usable signals from poorly reflecting surfaces up to 1 m away. Typical spot sizes are only 15 μm in diameter with standard lens and down to 1 μm through high-magnification microscope objectives. Fiber-optic interferometer simplifies vibration measurement where physical access is difficult, close stand-off-distances are required, and when relative motions between two points need to be measured.

Vibrometer controller OFV-5000. This modular high-end controller is used for the demodulation of velocity and displacement signals. The controller provides signals and power for the fiber-optic interferometer OFV-512, and processes the vibration signals. These are electronically converted by means of specially developed decoders installed within the controller in order to obtain velocity and displacement information about the test structure. Thus, the vibrometer controller outputs a modulated voltage in analog form corresponding directly to the vibrational velocity or displacement of an object, depending on the mode chosen by the operator. Current controller is equipped with wide-bandwidth velocity decoder VD-02 with best velocity resolution of 0.15 μm/s, maximum detectable velocity and frequency of 10 m/s and 1.5 MHz respectively.

Scanner controller (junction box) MSA-E-400 connects vibrometer controller and data management system and provides piezodriver for micro-scanning unit and amplifier for excitation signals.

Microscope with adapter MSV-100 and digital camera. Experimental setup is equipped with Nikon microscope, which is based on modular focusing unit MFU-4 that allows both coarse and fine focus adjustment. The microscope has got a set of objectives of different magnification power (up to 100×) and it is used to connect adapter MSV-100, which consists of dual beam-splitter module OFV-072 and micro-scanning module OFV-073. OFV-072 is mounted onto the microscope via CCD port (C-mount) and the optical fiber from the interferometer is coupled via scanning module OFV-073 into the optical path of the microscope. Reflected light from the test structure is coupled back into the laser interferometer OFV-512 for heterodyne demodulation in the controller OFV-5000. Scanning module OFV-073 employs a junction box MSA-E-400 in order to control two ultra-precise piezo-stages for scanning laser beam through the microscope optics. This allows changing laser spot position on the test structure by means of computer mouse within specialized software. Digital progressive scan camera VCT-101 with 1.3 Mpixels is used to provide a live video image of the test object with the location of laser beam spot during the whole measurement.

Data management sub-system consists of a Windows® 2000 based personal computer placed inside industrial housing, hardware for scanner control, video and data acquisition, as well as the MSV-400 software suite.

Excitation sub-system includes arbitrary waveform generator, amplifier and multi-layered piezoactuators for excitation of substrate with microdevices on it. Piezoelectric transducers were chosen due to their ability to rapidly and accurately expand and contract with the input signal at high frequencies.

The experimental setup described above was used to carry out modal analysis on the microcantilevers of the microswitches. The purpose of modal analysis was to measure natural frequencies of the microcantilevers and to determine their corresponding mode shapes. The measurements were performed by mechanically exciting microcantilevers with a frequency sweep actuation and determining the frequency response. In order to accomplish this, the substrate with microswitches was attached to a small piezoactuator and positioned under the microscope. The piezoelectric transducer was supplied with a broadband (100 ÷ 2,000 kHz) periodic chirp signal from a function generator for the excitation of all frequencies with the same energy. Before performing measurements, MSV-400 software was used for specification of a grid of scan-points on a live video image of tested microswitch at which system response will be detected afterwards. In addition, a proper alignment of the laser beam was carried out by means of microscope adapter unit MSV-100 in order to focus laser spot on a microcantilever and ensure that sufficient signal is being received by the LDV system. During the measurement procedure laser beam was continuously scanning (with the speed of 50 points/s) the surface of a microcantilever through the pre-defined points and registering their motion. Subsequently, the detected response signals in the time domain were transformed into the frequency domain using FFT algorithm and further processed by the data acquisition system within the computer workstation. As a result of these measurements and data processing several natural frequencies and mode shapes were determined for microcantilevers of different length. Table 5.4 lists the values of natural frequencies that were determined during this experimental modal analysis.

Meanwhile, Figs. 5.29 and 5.30 present 2-D and 3-D visualizations of obtained mode shapes. Actually, collected measurement data was used by software of the Polytec LDV system in order to generate animations of the vibration modes and the provided figures are just captured images from these animated results. This experimental data indicates that measured natural frequencies are essentially

Table 5.4 Values of natural frequencies determined from experimental modal analysis performed with microscopic laser Doppler vibrometry system MSV-400 from Polytec GmbH

Length of the microcantilever (μm)	Experimental value of natural frequency (kHz)
67	$f_1 = 1{,}540$ (flexural mode)
87	$f_1 = 978$ (flexural mode)
117	$f_1 = 853$ (flexural mode)
	$f_2 = 1{,}528$ (torsional mode)
	$f_3 = 1{,}702$ (flexural mode)

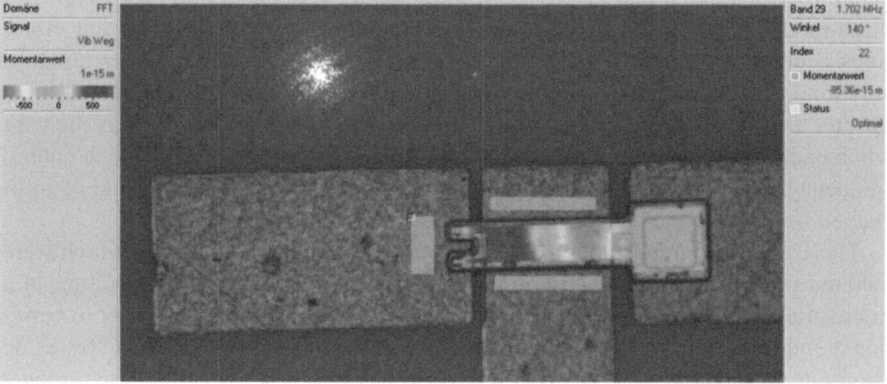

Fig. 5.29 2-D visualization of the measured mode shape corresponding to the second natural frequency of transverse vibrations of the microcantilever of $l = 117$ μm. The visualization of the mode shape is overlaid over the live video image of tested microswitch

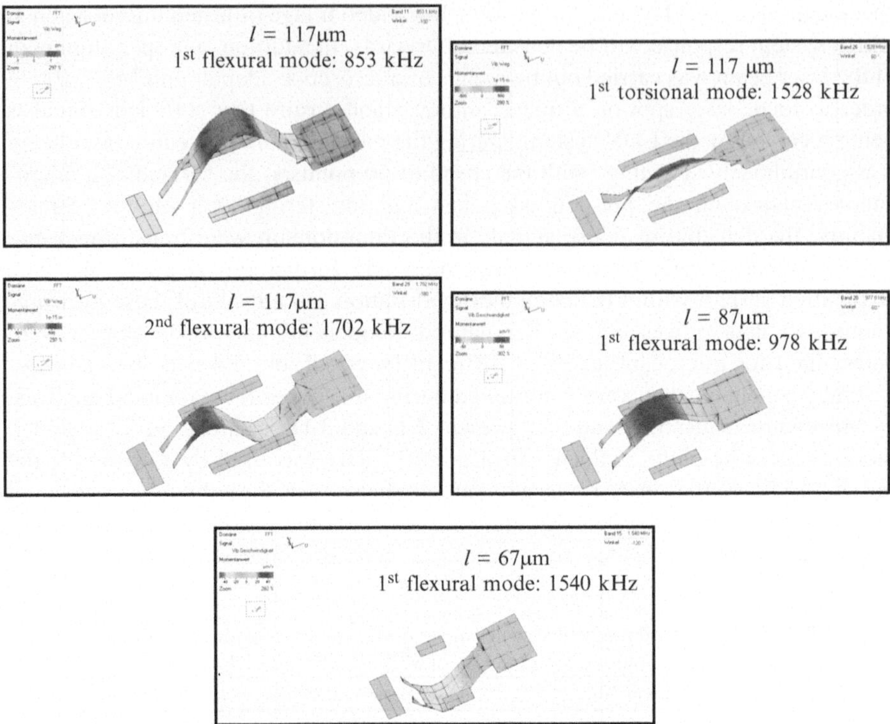

Fig. 5.30 Experimental natural frequencies and mode shapes of microcantilevers of different length

different from values predicted from finite element simulations for microcantilevers (Table 2.2). The reason for this discrepancy is particularly clearly visible from 3-D version of animated mode shapes of the microcantilever (Fig. 5.30). Examination of these mode shapes reveals that tested microcantilevers undergoing natural vibrations acquire deflection shapes that are characteristic not to cantilever but to a structure with fixed–fixed boundary conditions. The animations demonstrate that the free end of vibrating microcantilevers is fixed. The only exception is the case when microcantilever of $l = 117$ μm is vibrating in its first torsional mode at natural frequency of 1,528 kHz because here it is possible to notice that one side of fork-shaped free end of the microcantilever is moving freely.

These findings suggest that tested microcantilevers had contact tips which were adhered permanently to the drain electrode due to stiction thereby resulting in a microstructure with complicated fixed–fixed boundary conditions. In order to verify this assumption, numerical modal analysis was performed in COMSOL by using two finite element models of actual microcantilever of $l = 117$ μm: the first model had both ends fixed (Fig. 5.31 (top)), while the second model had one end fixed and the other end fixed in places of contact tips (Fig. 5.31 (bottom)). The latter model represented the assumed actual scenario of tested microcantilevers with tips

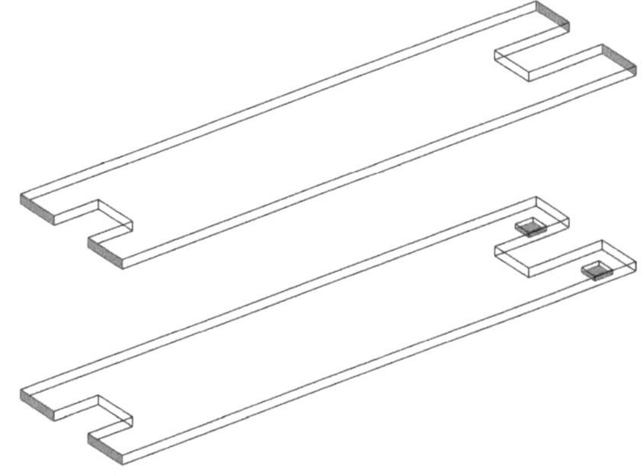

Fig. 5.31 Geometrical COMSOL models of a microstructure with different fixed–fixed boundary conditions. *Top*: both ends are fixed. *Bottom*: left end is fixed, right end is fixed in the area of two contact tips (fixation places are marked by a darker color)

Table 5.5 Comparison of natural frequencies for microcantilever of $l = 117$ μm, determined from experimental modal analysis and COMSOL simulations with models having two different fixed–fixed-type boundary conditions (Fig. 5.31)

Natural frequency (mode)	Experimental values, kHz	Simulated values (model 1 - Fig. 5.31(top)), kHz	Simulated values (model 2 - Fig. 5.31 (bottom)), kHz
First (flexural)	853	666	732
Second (torsional)	1,528	1,866	1,995
Third (flexural)	1,702	1,929	2,115

adhered to the bottom electrode. Simulations demonstrated that natural frequencies of the second model were higher than the values from the first model. Except for the first natural frequency, results from the first model were closer to the experimental values (Table 5.5). This inconsistency of experimental and simulation results leads to the conclusion that each tested microcantilever had a special case of adhered free end leading to the individual complicated non-ideal fixed–fixed boundary conditions. It is possible to identify at least three different scenarios of stiction of the free end of the microcantilever: (a) just one contact tip adhered to the bottom electrode, (b) both contact tips adhered, (c) free end of the microcantilever adhered to the bottom electrode not only in the region of contact tips but over a larger area.

References

1. Bagdonas V, Ostasevicius V (2009) Case study of microelectrostatic motor. In: Solid state phenomena: mechatronics systems and materials III: selected, peer reviewed papers from the 4th international conference: mechatronics systems and materials (MSM 2008), Bialystok, Poland, 14–17 July 2008, pp 113–118

2. Bilal S-N, Bagdonas V, Ostasevicius V (2008) Modeling dynamics of micromotor in viscous medium. In: Proceedings of Mechanika 2008: 13th international conference, Kaunas, Lithuania, 3–4 Apr 2008, pp 467–471
3. Bagdonas V, Ostasevicius V (2007) Design considerations of a microelectrostatic motor. J Vibroeng 9(4):55–59
4. Bagdonas V, Ostasevicius V (2007) Mathematical modeling and experimental analysis of electrostatic micromotor. In: Proceedings of Mechanika 2007: 12th international conference, Kaunas, Lithuania, 5 Apr 2007, pp 20–23
5. Ostasevicius V, Bagdonas V, Tamulevicius S, Grigaliunas V (2006) Analysis of a microelectrostatic motor. Solid State Phenom Mechatronic Syst Mater :185–189
6. Wu D, Fang N, Sun C (2006) Stiction problems in releasing of 3D microstructures and its solution. Sensor Actuat A Phys 128:109–115
7. Ostasevicius V, Dauksevicius R, Tamulevicius S, Bubulis A, Grigaliunas V, Palevicius A (2005) Design, fabrication and simulation of cantilever-type electrostatic micromechanical switch. In: Proceedings of SPIE: smart structures and materials 2005: smart electronics, MEMS, Bio-MEMS, and Nanotechnology, vol. 5763, 7–10 Mar 2005, San Diego, CA, pp 436–445
8. Ostasevicius V, Grigaliunas V, Tamulevicius S, Dauksevicius R (2005) A method for manufacturing of microelectromechanical switch. Patent No. LT5208B, 2005.04.25, The State Patent Bureau of the Republic of Lithuania
9. Vibrometer University (2007) In http://www.polytec.com/usa/158_463.asp. Cited 5 July 2007

The manufacturer's authorised representative in the EU is Springer Nature Customer Service Centre GmbH, Europaplatz 3, 69115 Heidelberg, Germany. If you have any concerns regarding our products, please contact ProductSafety@springernature.com

Printed and bound by CPI Group (UK) Ltd, Croydon, CR0 4YY
15/07/2026
02167619-0003